ORDINARY STATISTICS

SECOND EDITION

by

JAMES A. WALKER M.A. M.Ed

Dollar Academy

and

MARGARET M. McLEAN B.Sc

Stranraer Academy

EDWARD ARNOLD

First published 1973
by Edward Arnold [Publishers] Ltd
41 Bedford Square
London WC1B 3DQ

Edward Arnold (Australia) Pty Ltd
80 Waverley Road
Caulfield East
Victoria 3145
Australia

Reprinted 1975, 1976, 1978, 1979, 1980, 1981, 1982
Revised edition 1983
Reprinted 1984, 1985, 1986

ISBN 0 7131 0871 1

Acknowledgement
The tables in Appendix 1 first
appeared in *Three Figure Tables
for Maths and Science*, and are
reproduced by kind permission
of Blackie & Son Ltd.

Printed at The Bath Press, Avon

TO THE TEACHER

Statistics in the form of graphs, figures, opinion polls etc. are playing an increasingly large part in our everyday life. It is most important that children leaving school should have some knowledge of how this information is collected, what the various types of graph mean and above all how reliable the information is likely to be. A critical and perhaps even cynical view of the data presented by the popular press, T.V., advertisements and so on, should be fostered.

The purpose of this book is to introduce the basic ideas of Statistics and Probability in a modern way. It is intended to be a complete course in Statistics from first year up to GCSE. The subject matter covers both the English Associated Examining Board syllabus and the Scottish syllabus. These are far from being identical and so teachers should omit the sections or chapters not required for their own students; this may be done without any loss of continuity.

Some rather complicated mathematics has been included for any Sixth Form students who may be interested to pursue the subject more deeply. Students following a purely GCSE course would be well advised to omit the following sections which are not essential to the development of the subject—calculation of the standard deviation from an assumed mean, equation of the regression line, the proof of the Binomial Theorem.

Unfortunately, in some parts of the work there is bound to be some sheer, arithmetical "slog". This is unavoidable by the very nature of the subject. To minimise this as much as possible, the use of such aids as calculating machines, slide rules, mathematical tables for square roots etc. should be encouraged, provided they are available, of course.

As much as possible, the students should collect their own data, decide how to present it graphically and how to interpret what they have found out.

Three figure mathematical tables have been used for all the calculations.

In Statistics textbooks, we find a number of different notations used for the arithmetic mean, standard deviation etc. The notation and formulae used in this book are shown on the next page.

NOTATION AND FORMULAE

$N, X, \overline{X}, s$ refer to samples and n, x, μ, σ to populations

Arithmetic Mean:

$$\overline{X} = \frac{\Sigma X}{N} \quad \text{or} \quad \frac{\Sigma fX}{\Sigma f}$$

$$\mu = \frac{\Sigma x}{n} \quad \text{or} \quad \frac{\Sigma fx}{\Sigma f}$$

Standard Deviation:

$$s = \sqrt{\frac{\Sigma(X-\overline{X})^2}{N}} \quad \text{or} \quad \sqrt{\frac{\Sigma f(X-\overline{X})^2}{\Sigma f}}$$

$$\sigma = \sqrt{\frac{\Sigma(x-\mu)^2}{n}} \quad \text{or} \quad \sqrt{\frac{\Sigma f(x-\mu)^2}{\Sigma f}}$$

Standard Variable:

$$z = \frac{X-\overline{X}}{s} \quad \text{or} \quad \frac{x-\mu}{\sigma}$$

Binomial Distribution:

$$\mu = Np; \quad \sigma = \sqrt{(Npq)}$$

Samples of Size N (N large) from an infinite population:

$$\mu_{\overline{x}} = \mu; \quad \sigma_{\overline{x}} = \frac{\sigma}{\sqrt{N}}$$

INTRODUCTION

Statistics is the name given to the science of collecting facts and studying or analysing them. The facts or "data" can cover a tremendous range of subjects. They may be quite simple facts like how many bottles of milk a family buys every week or the goal average of a favourite football player. On the other hand, they may be extremely complex facts of the world of industry or nuclear physics and which can only be analysed by using computers.

If the data available on a subject is incomplete the answers we get to problems will only *probably* be true and are sometimes called predictions. An example of this is found in weather forecasting. Information about temperatures, wind, rain, barometric pressure etc. is collected in weather stations all over the world, on land and sea and even high in the air by means of weather balloons. When the weatherman has studied all this data, he has to predict what is *likely* to happen and, of course, he knows he is taking the risk of being wrong in his forecast.

Although statistics are becoming more and more widely used these days, their use is not a modern science by any means. For centuries men have collected information of various kinds and then put it to use. The Romans, for instance, conducted accurate censuses to count the number of people in the countries they conquered. It tells us in the Bible that Jesus was born in Bethlehem because Mary and Joseph had to travel there for the Roman Census. The Romans required that, on a certain date, every man (with his family) had to return to his birthplace to be counted. The purpose of the Roman Census was mainly to count the people who should have been paying taxes.

In this country, a national census is taken every ten years. Every household in the land is given a form with certain questions on it. This form must be filled in with information about everyone in the house at a certain hour on a certain date. People who are travelling at this time, are included in the census form at their destination whether it be a household or hotel. Institutions like jails, hospitals, boarding schools and so on also complete census forms. In this way almost every single person in the country (apart from tramps out on the road) is accounted for. The questions on the census form vary, to some extent, from one census to another depending on what facts the Government wants to know. Anyone who refuses to complete a census form is taken to court and fined or jailed.

Conducting a census is a mammoth and very expensive task. It cost the Government £3 million for the 1961 census. Even the distribution of forms involves great organisation. This is done by the Registrars of Births, Deaths and Marriages of each district in the country. The Registrars sub-divide their districts into small areas and employ a responsible person, over 21 years old, to distribute the forms in each area. This person must visit each house on his list and personally hand the form to an adult explaining what it is and that it must be filled in. As you may imagine, many visits will have to be made to some houses before anyone is found at home. After the census date, the forms have to be collected in the same way. Then all the forms are gathered together and the facts from all the forms in the country have to be tabulated. It is usually a year or two after a census before the results of it are ready to be published. Obviously a census of this comprehensive nature gives the Government excellent data about the population.

Statistics are all around us today. Whenever we open a newspaper we see facts and figures about something—what the "average" man spends on drink and tobacco; unemployment figures; road accident figures; business conditions; opinion polls. Often these facts are accompanied by a graph, diagram or chart of some kind. When we turn to the sports pages of the paper, we find them full of statistics about football, cricket, golf, etc.

Advertising is full of "statistics" too: '4 out of 5 housewives can't tell margarine from butter! 9 out of 10 people prefer "Sparkle" for washing up!' The advertisers, however, never tell us how they arrived at these figures.

In science and medicine, statistics are vitally important. Doctors, researching into new medicines and new treatments for diseases, very carefully collect and study the data from their experiments. In this way, they discover how effective the medicine or treatment may be. New drugs must be exhaustively tested for several years before they are released for general use.

Manufacturers and business men spend hundreds of thousands of pounds in building laboratories and employing scientists and mathematicians to conduct research into a great variety of products like modern building materials, new types of car engines, chemicals, new synthetic fibres and materials, etc. A great deal of this research is concerned with collecting data and drawing conclusions from it. If the results are not what is wanted then a new batch of experiments is set up, more data is collected and further conclusions drawn. This process may have to be repeated for months or even years, before the scientists achieve what they are after.

As you see, Statistics is part of every aspect of our daily life and in this book we shall be looking at how data can be collected to give as accurate a picture as possible, and seeing how we can interpret this data.

CONTENTS

THE COLLECTION AND REPRESENTATION OF DATA

Collection of data

This aspect of Statistics may, at first glance, appear to present no problems but in fact one of the most difficult tasks of a statistician is collecting accurate data.

Of course some facts are very easy to collect. For instance, there can be no doubt about the number of goals scored by football teams during a season as they are recorded by the Football League. On the other hand, if you ask a housewife how many loaves of bread she had bought every week for the last month, it is very unlikely that she would be able to tell you accurately.

There are different types of data which we may wish to collect—there are 'measurable' things like heights, weights; 'countable' things like the number of children in a family, the number of deaths due to motor accidents etc., and there is the information obtained by asking people questions. Opinion polls—very often on political issues—give the results of the latter.

The difficulties arise when we start to ask people questions.

The most reliable way of asking questions is by a questionnaire. Once a questionnaire has been compiled it has to be decided whether it should be sent through the post or whether an interviewer should be used to read aloud the questions and record the answers. The second method is much better. It ensures that each question is answered, and even more important, that each question is understood. The trouble with sending

questionnaires by post, to be answered by the people concerned on their own, is that often the questions are not understood and even more often the questionnaires are never filled in and returned. So if you wish to collect information by sending out questionnaires by post you can only rely on a very small percentage being returned.

Having considered how to ask the questions, we must think very carefully of *what* questions are to be asked. They must be very carefully worded so that as far as possible there can be no doubt about their meaning. For instance, if we ask: 'How many good meals did you have last week?' the person questioned has to decide for himself exactly what is meant by a good meal—does it mean a substantial three course meal or does it mean a well cooked meal? Also, people's ideas will vary about what constitutes a substantial meal or a well cooked one.

We must also recognise the fact that on being questioned people either intentionally or subconsciously may be dishonest. If we ask a man in the presence of his wife how much he spends on gambling every week, we may receive an answer very different from one obtained from him privately! Also in a sense trying to please you or to put themselves in a good light, people often give you the answer they think is *expected*. The person being questioned should never be told what the survey is about. This also helps to avoid being given the 'right' answers.

1

A series of questions which builds up the required answer may be formulated. These may be self-checking in some way. For example, the same questions may be put in several ways and any inconsistency in the replies is revealed. If any inconsistency is very marked then that particular questionnaire must be scrapped.

Suppose you wish to conduct a survey of how many people have been seriously ill over the past year. The bald question would be: 'Have you been seriously ill over the past year?' Various types of responses are likely to be forthcoming. Some people would be unwilling to admit, even to themselves, that they had been very ill; some people would not know how ill they had been and some would be inclined to exaggerate a minor illness. You would, of course, get a number of perfectly honest replies—but which ones and how many? In a case like this, a series of 'leading questions' can build up a correct picture for the investigator.

1. Did you attend the doctor last year?
2. Were you attending the doctor regularly?
3. Were you confined to bed for more than a day or two?
4. Did you require to go to hospital?
5. How long were you in hospital?
6. What kind of treatment did you receive —an operation or medication etc.?
7. How long a convalescence did you have?

These 'small' questions are more likely to be answered correctly, since the person being interviewed usually does not sense the drift of the questions and so give a 'loaded' answer.

In designing a successful questionnaire, certain simple rules must be adhered to:

1. The questions must be very simply worded and straightforward.

2. They should be of such a nature that the answers are 'yes' or 'no', or a precise answer such as a number or place.

3. There must be no ambiguity—only one interpretation should be possible.

4. Any question of a personal nature should be very tactful.

Having now considered how to ask questions and what kind of questions to ask, the remaining problem is whom you ask. This is the most difficult problem to solve.

If some information is wanted about a particular small group of people, then each person in the group can be questioned individually. If, however, you want information about a large number of people, say the population of a large city or indeed the population of Great Britain, then it is impossible to question every one of them (except, of course, in a Government Census). In this situation, and this is the most usual one, we must question a *sample* of the population.

How do we choose a sample which is representative of the whole group? If you pick out every hundredth name in a telephone directory, you are not considering people who have no telephones. This would give a *biased* sample. If you stopped every fiftieth person in a street near a factory at five o'clock, then your sample would be biased towards factory workers. If you stopped every fiftieth person at ten thirty a.m. near a busy shopping centre, then your sample would be biased towards housewives.

You may, of course, *want* a sample which is biased towards a certain group. If a market research firm is dealing with the effect of advertising on sales of soap powder, then the sample should be biased towards the group of people who usually choose the brand of soap powder that a family buys, namely the housewives.

One way of obtaining a reasonably unbiased sample is to use the electoral roll of the town and go through it picking out every fiftieth or hundredth name (depending on how many people are to be in the sample). This, however, does not take into account people under the age of eighteen years, a factor that could be allowed for by making sure that a number of young people were also questioned.

Another kind of sample to take is a *random sample*; this is dealt with in the chapter on Sampling. There is a rather complicated way of obtaining a really *representative* sample of a given population, which is used by good market research firms. but it involves a great deal of work and is outside the scope of this book.

The purpose of choosing a sample need not be for the answering of questions, it may be for finding out other information such as weights, or heights of the members of the sample.

Representation of data

Once we have collected our data, we need to present it in an attractive and eye-catching manner. The obvious way of doing this is by means of some kind of graph or chart. These graphs can be made attractive by the use of the various types of modern colouring pens—felt tipped and fibre tipped pens etc. The colours are varied and very bright—much better than colouring pencils, but much more expensive.

We often see graphs of different kinds in newspapers, magazines etc. Start a collection of these graphs. By the time you have

finished this chaper, you should be able to decide whether the graphs you have collected are good representations of the data or misleading ones.

Pictographs

One of the most arresting ways of illustrating statistics is by using a graph in the form of pictures. This kind of graph is called a **pictograph** (sometimes **pictogram).** These pictures may be of cars, houses, milk bottles, aeroplanes, people etc., reduced in size but drawn to a definite scale for purposes of comparison.

Local councils or corporations often use some kind of pictograph to show how the money they take in from the rates is used. Each council must produce a balance sheet every year to inform the ratepayers how their money has been spent. Many councils put a great deal of effort into showing this information in the form of a pictograph. The reason for this, is that the information is more easily seen and appreciated by people when presented by pictures, than in a list of facts and figures.

Let us imagine that a certain local council presented the following balance sheet (page 4) for the ratepayers one year.

Fig. 1.1

FOR EACH £

32½p
Education

13½p
Cleansing

11p
Police

9½p
Roads

9p
Health & Welfare

5p
Lighting

4½p
Housing

4p
Parks

3½p
Fire Service

3p
Libraries

2½p
Water

2p
Miscellaneous

For each £ spent:		
Education	cost	$32\frac{1}{2}$p
Cleansing	,,	$13\frac{1}{2}$p
Police	,,	11p
Roads	,,	$9\frac{1}{2}$p
Health and Welfare	,,	9p
Lighting	,,	5p
Housing	,,	$4\frac{1}{2}$p
Parks	,,	4p
Fire Service	,,	$3\frac{1}{2}$p
Libraries	,,	3p
Water	,,	$2\frac{1}{2}$p
Miscellaneous	,,	2p

Figure 1.1 (page 3) shows one way of illustrating these figures in a pictograph. The 'money bags' are drawn in proportion to the amounts of money (in respect of length).

Make up a pictograph of your own to illustrate the same figures. Then compare your pictographs.

Try to obtain pictographs of this kind put out by local councils (try the local public library) and study them.

Figure 1.2 shows the number of cases of tinned peaches sold by four shops in a month. Study the pictograph and see if you think this is an accurate way to represent the data. If you think it is not, give reasons for your answer. This pictograph reveals clearly the weakness of this kind of graph.

Shop A sold twice as many cases as Shop B and so the cube for Shop A is of side 4 cm while the cube for Shop B is of side 2 cm i.e. the dimensions for the first cube

Represents 10

Shop A   Shop B   Shop C   Shop D

Fig. 1.3

are twice what they are for the second *but* the resulting picture produces the impression of volume and Shop A has *apparently* sold a great deal more than twice what Shop B sold. How much more according to volume?

In drawing pictographs of this kind, even when keeping the linear dimensions in proportion to the data, the resulting pictures can be in a very different ratio.

Fig. 1.2

120   60   40   30

SHOP A   SHOP B   SHOP C   SHOP D

A much more accurate pictograph of the data is that shown in figure 1.3 opposite. There one small case is used to represent every ten cases.

EXERCISE A

1. Study this pictograph and answer the questions below.

Fig. 1.4 The Number of Hours of Bright Sunshine in Jan. 1965

Lerwick   Stranraer   Perth   Kirkcaldy

 Represents 10 hours

a) Why has this been drawn on squared paper?
b) What does one picture of the 'sun' represent?
c) How many hours of bright sunshine did each of the four places have?
d) What does half a 'sun' stand for?
e) What would a quarter of a 'sun' stand for?

2. In the same year the following hours of sunshine were also recorded in January.

Stornoway—36, Cardiff—32, Lowestoft—48, Paisley—22.

Draw a pictograph like the one in question 1. This time let a 'sun' represent eight hours of sunshine.
    What will half a 'sun' represent?
    What will quarter of a 'sun' represent?

3. Study the following pictograph and answer the questions. (Notice how the columns of 'fish' are evenly spaced out).

Fig. 1.5 The Catches of Five Anglers in a Season's Fishing

Mr.Black Mr.Green Mr.White Mr.Brown Mr.Gray

 Represents 20 fish

a) What is the scale of this pictograph?
b) Which two men had caught the same number of fish? How many did each catch?
c) Which man had a catch of 50 fish?
d) Which man had a catch of 220 fish?
e) What size of catch do you estimate Mr Gray had?
f) Which man won the fishing prize?
g) Who deserved the 'booby' prize?

4. The rainfall (in centimetres) recorded one year for these places was as follows:
Renfrew—90; Kinloss—25; Benbecula—110, Dunoon—210, Helensburgh—125.
Draw a pictograph to illustrate this using an

umbrella, ☂ to represent 10 cm of rain.
Watch these points—
    1. Make your umbrellas the same size.
    2. Space them out evenly.
    3. Have a title.
    4. Show your scale.
    5. Put in the names.
(This pictograph could be made very attractive, by shading the tops of the umbrellas a different colour for each town). Dunoon is a popular holiday place. Would you choose it for a holiday?

5

5. The number of bottles of milk used in a small school for each week of a month are shown below.

1st week — 40 bottles
2nd week — 36 bottles
3rd week — 44 bottles
4th week — 38 bottles

Draw a pictograph to illustrate this, using a milk bottle for your picture. This time choose your own scale. Remember to watch for the five points mentioned in question 4.

6. A small poultry farmer with free range hens sold the following numbers of eggs each month over six months. Draw a graph to show this, choosing your own scale.

Jan. — 5 dozen   April — $14\frac{1}{2}$ dozen
Feb. — 4 dozen   May — 12 dozen
Mar. — 9 dozen   June — $10\frac{1}{2}$ dozen

Find out some data for yourselves and draw pictographs to show it.

Bar charts

The pictographs you were drawing and looking at in the last section are an attractive way of presenting data but they have two main drawbacks which you have probably found out for yourselves, namely,

1. They take a long time to draw well.
2. They are difficult to read accurately, when there are 'bits' of pictures.

If instead of a column of pictures, we simply draw a plain column, then we overcome both difficulties.

Figure 1.6 is a **bar chart** or **column graph** of the number of bottles of milk used in the school in question 5 of the last exercise.

Study the graph and note the following.

1. The title tells us what the graph is about.
2. The columns are all the same thickness and are spaced out evenly along the *horizontal axis*.
3. The *vertical axis* has a scale (like a ruler) which tells us the number of bottles.
4. The columns rise to the correct height for each week measured against the vertical scale.

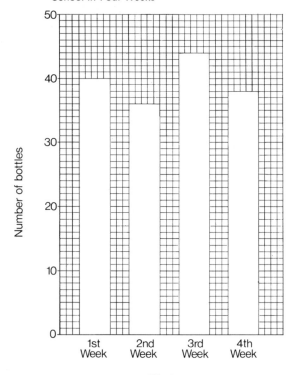

Fig. 1.6 The Number of Bottles of Milk Used in a Small School in Four Weeks

EXERCISE B

1. In a school one year, a second year class decided to investigate how all the pupils travelled to school. Figure 1.7 shows how the first year pupils travelled to school. Some of the pupils live in country areas well away from normal bus routes and these children are collected by taxis or by small vans used later in the day for transporting meals from central kitchens to the outlying primary schools.

a) What is the most usual method of travelling to school?
b) What is the least usual method of travelling to school?
c) What transport is used by 30 pupils?
d) How do 57 of the pupils travel?
e) How many cycle to and from school?
f) How many pupils are collected by taxis or meals vans?
g) How many first year pupils were present the day the survey was done?

6

Fig. 1.7 The Means of Travelling to School of First Year Pupils

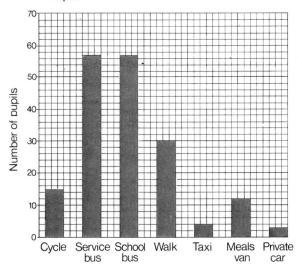

2. Conduct a similar survey either in your own class or in several classes and draw a graph of your results.

3. Study this graph and answer the questions.

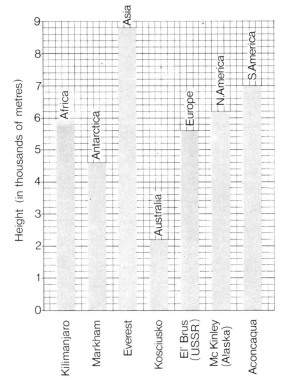

Fig. 1.8 The Highest Mountain in Each Continent

a) Which continent has the highest mountain in the world?
b) Which continent does not have any very high mountains?
c) What is the approximate height of Mount McKinley?
d) What mountain is approximately 7000 metres?
e) In which continent is there a mountain of about 5900 metres?
f) Which mountain is slightly more than half the height of Mount Everest?

4. Use your Atlas to find the heights of some of the highest mountains in Britain and draw a graph of the heights.

5. Study the graph and answer the following questions.

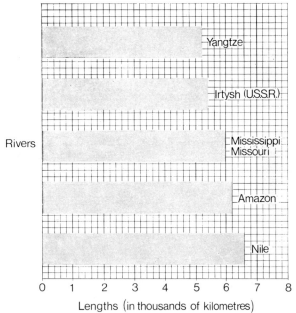

Fig. 1.9 Longest Rivers in the World

a) Why do you think the columns have been drawn horizontally for this graph?
b) What is the approximate length of the longest river in the world?
c) Which river is about 5400 kilometres long?
d) Which is about 6200 kilometres long?
e) Which two of the rivers shown are most nearly the same length?

7

6. Use your Atlas to find the lengths of the longest British rivers and draw a graph to illustrate the lengths, drawing your columns horizontally.

7. The absences shown in the graph below were recorded in a school during a 'flu epidemic in February 1968, for two consecutive school weeks. (Notice how we can use column graphs for purposes of comparison.)

a) Which of the two classes was affected first by the 'flu outbreak?

b) On how many days were there more than 10 pupils absent in class A?

c) On how many days were more than 10 pupils of class B absent?

d) What was the largest number of pupils to be absent on any day in either class?

e) Between what two days was there the biggest jump in the absences for (i) class A (ii) class B?

f) On what days were there more than eight pupils absent in both classes?

8. A pupil conducted a survey in two first year classes on their favourite flavours in potato crisps. The results are shown below. Draw a graph similar to the one in question 7 to illustrate this—choosing your own method of shading.

Crisps' Flavour	Class A	Class B
Smoky Bacon	6	14
Cheese and Onion	7	13
Plain	1	0
Salt and Vinegar	18	5

9. Draw a graph grouping the subjects for each boy together to illustrate the examination marks of these boys in the subjects shown.

Boy	English	Maths	Art
Alan	60	50	60
Bert	63	50	53
Colin	87	65	30
David	60	43	75

Fig. 1.10 Class Absences over Two Weeks During a Flu Epidemic

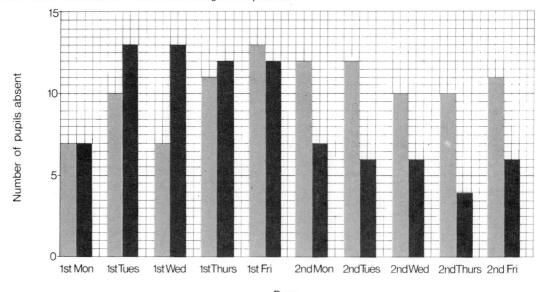

Class A   Class B

10. Draw a graph to show the medal positions of the top four countries after the Mexico Olympics in 1968.

Country	Gold	Silver	Bronze
United States	45	27	34
Soviet Union	29	32	30
Japan	11	7	7
Hungary	10	10	12

Collect data for yourselves in your class or in other classes and draw graphs; such things as hair colours, eye colours, shoe sizes, weights, heights, favourite pop groups, TV stars, TV programmes etc.

Composite bar charts

This is a type of bar chart which has each bar sub-divided into several sections and these sections are shaded to illustrate the data.

Example
The following chart shows the land use on three farms.

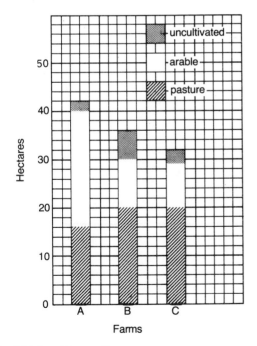

Fig. 1.11 Land Use on Farms

With this type of graph we can compare each farm one to another and also we can compare the land use within each individual farm.

EXERCISE C

1. Draw a composite bar chart to show the comparative medal positions for the four countries as given in question 10 of the last exercise.

2.

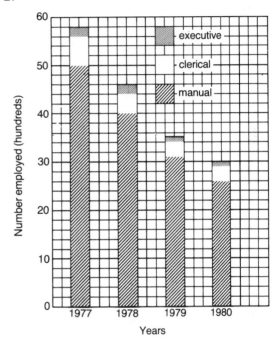

Fig. 1.12

This diagram shows the staff position of a firm over several years. Answer the following questions.
  a) What was the total staff in 1977?
  b) What was the total staff in 1980?
  c) By what percentage was the staff reduced between 1977 and 1980?
  d) What fraction of the total staff consisted of manual workers in 1978?
  e) What fraction of the total staff consisted of clerical workers in 1979?

3. Draw a composite bar chart to illustrate

the grades obtained by Higher Mathematics candidates in a Scottish school over a number of years.

Years	Grades		
	A	B	C
1978	7	14	20
1979	5	20	20
1980	3	14	22
1981	12	18	25

4. Draw a horizontal column graph to illustrate the following stopping distances for cars. By different shading differentiate between the 'thinking distance' and 'braking distance'. These stopping distances are for private cars or small vans with good drivers, perfect brakes, tyres etc. in broad daylight with good, dry roads. Larger vehicles under the same conditions may need twice these distances. On wet roads, for all vehicles, twice the normal distances are required.

Speed	Thinking Distance	Braking Distance	Overall Stopping Distance
km/h	metres	metres	metres
32	6	6	12
48	9	13.5	22.5
64	12	24	36
80	15	37.5	52.5
96	18	54	72

5. Illustrate the information given in question 7 of the last exercise in the form of a composite bar chart.
   a) On which day was there the greatest total number of pupils absent?
   b) Between which two successive days was there the greatest increase in the total number of pupils absent?
   c) Between which two successive days was there the greatest drop in the total number of pupils absent?

6. The distribution of full-time students at a university by faculties was as follows.

	1968/9	1975/6
Science	540	780
Engineering	1055	920
Social studies	460	520
Environmental studies	280	270
Postgraduate students	100	250

Illustrate the information in the form of a composite bar chart.

Line graphs

Line graphs as the name suggests show the data by means of drawing a line. This kind of graph is very good for showing upward or downward trends, and is the kind of graph most used and *misused* in newspapers, magazines and advertisements.

Example

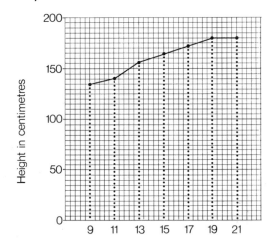

Fig. 1.13 A Boy's Height over Several Years

Notice how the points are plotted. A line is taken up from the ages to the required height. Then the points are joined with straight lines.

The boy grew most between the ages of 11 and 13 years and so the line has the steepest upward slope between those ages.

10

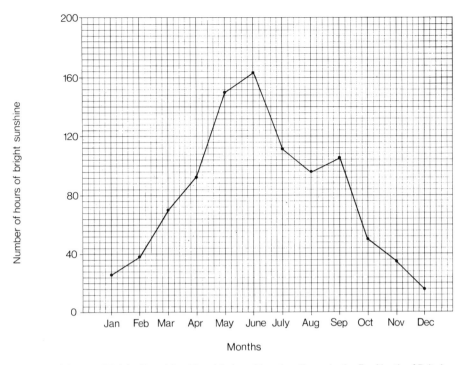

Fig. 1.14 The Number of Hours of Bright Sunshine Monthly for a Year in a Town in the Far North of Britain

Between the ages of 19 and 21 years, he did not grow at all and so the line is level.

EXERCISE D

1. a) Does the graph above (Figure 1.14) show what you would expect concerning the monthly number of hours of bright sunshine?

b) What two months had less sunshine than you might expect?

c) Between what two months was there the biggest increase in the number of hours of sunshine?

d) Two different periods show an almost identical drop in the number of hours of sunshine. What are they?

2. In the same year the number of hours of bright sunshine recorded in a town much further south were as follows.

Draw a graph similar to the one in question 1 using these figures (or try to find out the figures for your own town) and

Jan.	—	57 hours	July	— 188 hours
Feb.	—	84 ,,	Aug.	— 139 ,,
March	—	44 ,,	Sept.	— 146 ,,
April	—	136 ,,	Oct.	— 75 ,,
May	—	185 ,,	Nov.	— 71 ,,
June	—	209 ,,	Dec.	— 47 ,,

compare and contrast the two graphs.

3. The infant mortality rates (per 1000 births) in Glasgow are shown over a period of years.

1930–34 (average)	— 102
1935–39 ,,	— 93
1940–44 ,,	— 95
1945–49 ,,	— 64
1950–54 ,,	— 37
1955–59 ,,	— 35
1960–64 ,,	— 33

Draw a line graph to illustrate this and comment on it, giving reasons for the trend it shows.

Why should the infant mortality be given as a rate and not as the actual numbers of infants who died?

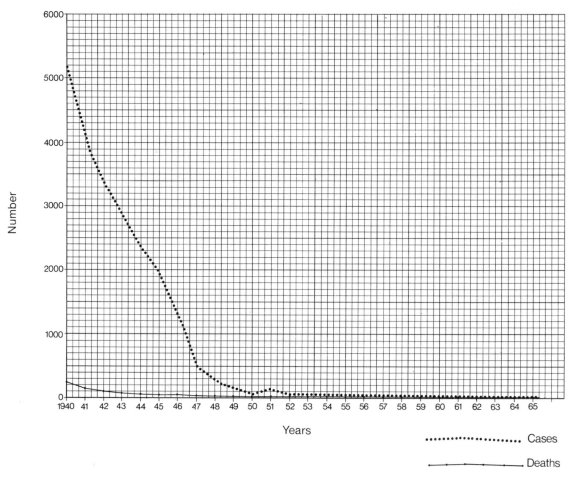

Fig. 1.15 The Incidence and Deaths From Diphtheria in a Large City (1940–65)

4. Find out the infant mortality rates for your own town or area, draw a line graph to illustrate them and then compare and contrast the two graphs.

(The figures for questions 3 and 4 may be found in the Registrar General's Annual Reports—separate reports being published for England and Wales, and Scotland.)

5. The graph above (Fig. 1.15) shows a very different trend. Study the graph and comment on the statement that it 'represents a lesson in the value of intensive preventive medicine.'

6. Draw a graph from the figures given below to show the comparison in the annual

DEATH-RATES (PER 100 000) FROM PULMONARY TUBERCULOSIS											
Year	1955	–56	–57	–58	–59	–60	–61	–62	–63	–64	–65
Glasgow	28	25	24	26	20	19	18	18	21	14	14
Aberdeen	8	10	5	7	6	5	5	2	4	1	3
Manchester	19	15	14	10	12	12	8	11	8	8	7

12

death-rate (per 100 000 population) from pulmonary tuberculosis in Glasgow, Aberdeen and Manchester from 1955–65. Comment on the trends shown and the differences between the three cities.

7. The annual incidence of whooping cough in a large city from 1940–65, is as follows.

Year		Cases	Deaths
Average 1940–44		4463	92
„	1945–49	3321	32
„	1950–54	4794	13
„	1955–59	2276	3
„	1960–64	1657	1
„	1960	3745	4
„	1961	824	—
„	1962	272	—
„	1963	2695	2
„	1964	751	—
„	1965	459	—

Draw line graphs to show the cases and the deaths on a *large* sheet of graph paper—being very careful of your time scale.

Comment on the trend shown in the graph.

Draw another graph to illustrate the same data as if you were trying to publicise the need for children to be immunised against whooping cough—the sort of graph you might see on a poster. Try to make up a bold, arresting title.

8. The Registrar General of Scotland, on the basis of all the information at his disposal about death-rates, birth-rates, emigration rates etc., in 1964 estimated the population of Scotland up to 1981 as shown below.

Year	Population (in thousands)	
	Male	*Female*
1964	2534	2708
1969	2576	2734
1974	2650	2789
1976	2687	2817
1981	2789	2898

Draw a graph to illustrate this showing males and females separately. (Watch your time scale.)

Comment on the trend.

Try to find out the latest figures for the population (try the local registrar of births, deaths and marriages) and see whether this prediction is proving accurate. If not try to find out the reason.

Misuse of line graphs

Line graphs often appear in newspapers, magazines and advertisements and can be very uninformative and even misleading once you take a close look at them.

The 'Ten Year Growth' of a company is shown in these figures published in a daily newspaper recently.

Year (ended 31st March)	Profit before Tax (£ millions)
1971	34
1970	30.9
1969	29.8
1968	27.7
1967	25
1966	22.5
1965	21.3
1964	20.5
1963	17.8
1962	15.1

Let us look at a few ways in which these figures could be shown graphically.

Business is doing very nicely.

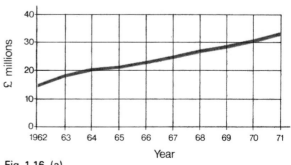

Fig. 1.16 (a)

13

Business is looking up.

Fig. 1.16 (b)

Business is booming!

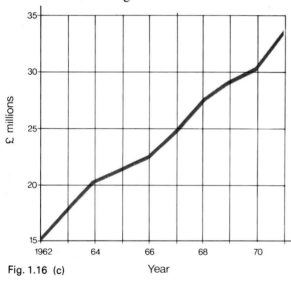

Fig. 1.16 (c)

As you can see, it is very easy to alter the 'look' of a line graph by adjusting one or both of the scales.

If you study newspaper graphs like the one below, you will notice that the vertical scale

Fig. 1.17

14

(often numbers or amounts of money) very rarely starts at zero. The horizontal scale—usually a time scale—can be squeezed closer or spread out to give very different impressions. Indeed, sometimes the time scale is not even regular.

As you may imagine, the resulting graph was quite misleading.

Some graphs, particularly in advertisements, use very vague terms. For instance, some advertisements for headache tablets or powders show graphs of the time it takes the tablets to overcome the 'threshold of pain'—making no attempt to define this term. Still—the graph looks impressive!

We must then, when we see graphs in newspapers etc., take more than just a cursory look (which creates the impression the graphs were intended to create); we must study them carefully, looking at the scales and at the terms used. Only then can we decide what value the graph has.

Pie charts

Another way of illustrating data is by means of a **pie chart.** This is a circle divided up into sections which are usually shaded in various ways.

Figure 1.18 shows a pie chart illustrating the eye colours of a class of 36 pupils.

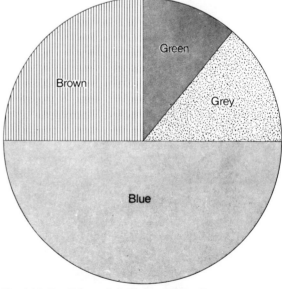

Fig. 1.18 Eye Colours in a Class of 36 Pupils

To read a pie chart you must first measure for each section the angle at the centre of the circle.

Thus, blue eyes has an angle of 180°
  brown ,, ,, ,, ,, ,, 90°
  grey ,, ,, ,, ,, ,, 50°
  green ,, ,, ,, ,, ,, 40°

The whole circle represents the whole class, and a fraction of the circle represents the same fraction of the class.

The fraction of the circle representing blue eyes

$$= \frac{180}{360} = \frac{1}{2}$$

Thus, the fraction of the class with blue eyes

$$= \frac{1}{2} \text{ of } 36$$

$$= 18$$

Similarly, the number of pupils with brown eyes

$$= \frac{90}{360} \text{ of } 36$$

$$= 9$$

the number of pupils with grey eyes

$$= \frac{50}{360} \text{ of } 36$$

$$= 5$$

the number of pupils with green eyes

$$= \frac{40}{360} \text{ of } 36$$

$$= 4$$

Example

Draw a pie chart to illustrate these road deaths in Scotland one year, for boys aged up to and including 19 years.

Pedestrians   = 79
Pedal cyclists   = 5
Drivers or passengers on motor cycles = 27
Drivers or passengers in motor cars = 46
    Total deaths = 157

Solution

First we must change the fraction of the total deaths for each type into the angle to be used in the pie chart.

Pedestrians $\frac{79}{157}$ of total deaths.

On the pie chart, the angle required

$$= \frac{79}{157} \times 360°$$

$$= 181.2° \text{ (approx)}.$$

Pedal cyclists $\frac{5}{157}$ of total deaths.

On pie chart, the angle required

$$= \frac{5}{157} \times 360°$$

$$= 11.5° \text{ (approx)}.$$

Drivers or passengers on motor cycles

$$= \frac{27}{157} \text{ of total deaths}$$

On the pie chart, the angle required

$$= \frac{27}{157} \times 360°$$

$$= 61.9° \text{ (approx)}.$$

Drivers or passengers in motor cars

$$= \frac{46}{157} \text{ of total deaths}$$

On the pie chart, the angle required

$$= \frac{46}{157} \times 360°$$

$$= 105.5° \text{ (approx)}.$$

Fig. 1.19 Road Deaths for a Year in Scotland (Boys aged up to and including 19 Years)

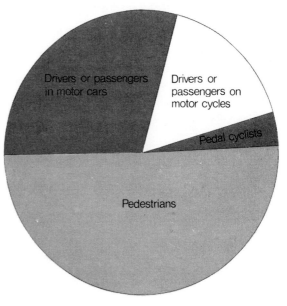

1. Measure the angles of Fig. 1.20 and calculate the number of pupils with each colour of hair.

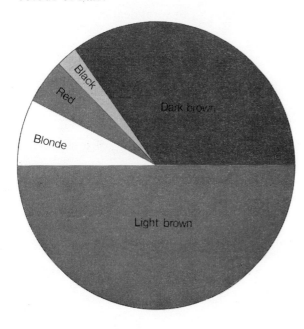

Fig. 1.20 Hair Colours in a Class of 40 Pupils

2. This data shows the way in which the world's catch of fish is composed of the various types of fish. Draw a pie chart to illustrate this.

33% Herring family (herrings, pilchards, sprats, sardines, shad, white bait and anchovies)

15% Demersal fish (the white fish—cod, haddock, whiting, ling, hake, and the flat fish—halibut, flounder, plaice, sole.)

31% Other Marine Species (including tuna, barracudas, mullets, perches, sharks etc.)

11% Freshwater Species

2% Salmon

8% Crustaceans and Molluscs (crustaceans—lobsters, shrimps etc. molluscs—oysters, mussels etc.)

3. Draw a pie chart to illustrate the composition of the earth's atmosphere.

Nitrogen 78%
Oxygen 21%
Argon almost 1%
(very small traces of other gases)

Collect data of your own from your class or another class and draw pie charts to illustrate.

Disadvantages of pie charts

Pie charts have certain disadvantages.
1. There has to be actual measurements of angles using protractors.
2. There is quite a bit of calculation involved.
3. The answers in most cases must be approximations.

However, pie charts have one **advantage**—they show how one whole thing is divided into parts and what size these parts are in relation to each other and to the whole.

EXERCISE F Miscellaneous

1.

Pupil	Mark	
	1st Exam.	2nd Exam.
Anne	79	82
Joan	52	71
Diana	80	65
Mary	24	50
Yvonne	62	50

This table shows the marks of five girls in two arithmetic examinations. Draw a bar graph to illustrate these results.

a) Which girl improved her performance most in the second examination?

b) Which girl had the poorest result over the two examinations?

c) Which girls improved their marks in the second examination?

2. Draw a line graph comparing the profits of Company A and Company B over the years 1974 to 1980.

Year	Profits (£ thousands)	
	Company A	Company B
1974	25	25
1975	29	30
1976	34	25
1977	46	23
1978	70	15
1979	82	15
1980	93	10

Country	Class A	Class B
France	7	5
Germany	6	5
Switzerland	7	6
Spain	6	5
Norway	2	4
Greece	1	2
Italy	1	3
	30	30

 3.

 5.

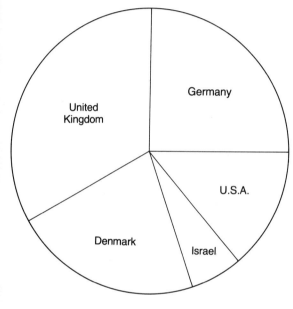

Fig. 1.21 Sales Figures of a Company in Various Countries

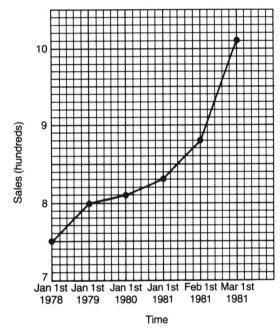

Fig. 1.22 Monthly Average Sales Figures for a Small Factory

This pie chart shows the sales figures of a company in various countries.

If the sales in the U.K. amounted to £2 400 000, calculate the total sales figure and the sales figures for each of the other countries.

4. A survey was conducted in two classes in a school asking which European country the pupils would most like to visit. The results are shown below. Illustrate these results
    a) in a compound bar graph,
    b) in two pie charts.

This graph shows the monthly average sales figures for a small factory producing machine parts. Give two reasons as to why this graph could be deceptive to someone not looking closely at it.

6. The pictograph at the top of p. 18 shows the sales of butter in a small shop over four weeks. Comment on why the diagram may be rather misleading. Draw a pictograph which you think would illustrate the data more accurately.

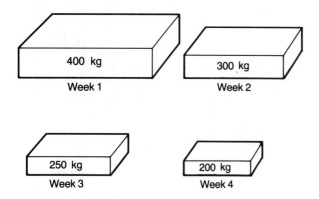

Fig. 1.23 Butter Sales in a Small Shop over Four Weeks

7. The following graph shows the profits of a certain company over a number of years. The graph shows the company in a favourable light to potential investors. Comment on two things you think could be misleading. Draw a graph of your own which you think would be more accurate.

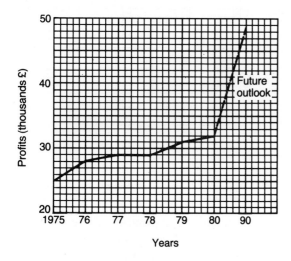

Fig. 1.24 Profits of a Company Since 1975

8. The pictograph shown illustrates the daily average number of hours of sunshine in a holiday resort over three consecutive weeks. Comment on why this graph might be misleading. Draw a pictograph to illustrate the data in what you think would be a better way.

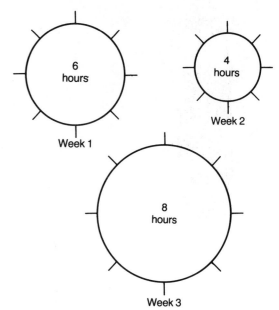

Fig. 1.25 Daily Average Sunshine Hours over Three Weeks

9. The following graph shows the profits of a company for the years 1970–77. Describe two ways in which this graph is deceptive. Draw a line graph illustrating the same data in what you think is a more accurate manner.

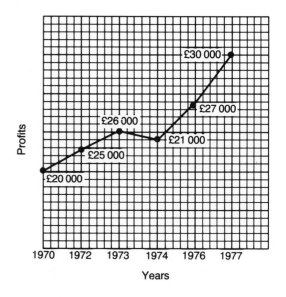

Fig. 1.26 Company Profits for 1970–77

18

2 ESTIMATION OF ERROR

Since much of the data that is considered in Statistics is obtained by measurement, it is important that we should understand how accurate our information is and based on this information, how accurate any further calculations will be.

It seems appropriate, then, at this stage in the course to consider how we may estimate the extent of any error that may be contained in our data.

Approximations

Significant figures

58.4819 may be written as

58.482 correct to 5 significant figures
58.48 correct to 4 significant figures
58.5   correct to 3 significant figures
58   correct to 2 significant figures
60   correct to 1 significant figure

Decimal places

58.4819 may be written as

58.482 correct to 3 decimal places
58.48 correct to 2 decimal places
58.5   correct to 1 decimal place
58   correct to the nearest whole number
60   correct to the nearest ten

EXERCISE A

1. Express the following, correct to the number of significant figures shown in the brackets.

   a) 5.632 (3)   *b*) 10.08 (3)   *c*) 10.03 (2)
   d) 0.00681 (2) *e*) 6.138 (1)   *f*) 117.2 (2)

2. Express correct to 1 decimal place.

   a) 11.66   *b*) 9.81   *c*) 69.008
   d) 0.085   *e*) 6.99   *f*) 15.849
   g) 15.850 *h*) 15.851

3. Express 582 733

   a) to the nearest 100 000
   b) to the nearest 10 000
   c) to the nearest   1 000
   d) to the nearest   100
   e) to the nearest   10

Absolute error

In dealing with certain data the information is found by counting and there can only be one correct answer e.g. the number of pupils on the roll of a particular school, the number of runs scored by a cricket side during a season or the number of goals scored by a hockey eleven during a winter. Other information is found by measuring, and this is a very different situation.

In measuring anything whether it be the length of a line, a period of time or the mass of an object we are limited in our accuracy by the equipment available and our own human limitations. We can never find the *exact* measure we are seeking and we must be content with an approximation. The difference between the true measure and that obtained by measurement is called the **error** (though there has been no mistake in the measuring). This error can be reduced, of course, by using more accurate instruments but a measurement can never be exact and so errors can never be eliminated completely.

It is important then that we should be aware of what error is implied by our measurements and what the maximum possible error is likely to be.

Let us consider the measurement of a straight line. Using an ordinary ruler the smallest unit of measurement that we can read off, with any accuracy is a millimetre or 0.1 cm. If our line is measured as 4.9 cm, this means that it measured 4.9 cm *to the nearest* 0.1 cm i.e. the line was *actually* between 4.85 cm and 4.95 cm long. The maximum error then is 0.05 above our nominal value and 0.05 below it. This is called the **absolute error** (Note that the absolute error is half the smallest unit of measurement. In the example above, the smallest unit of measurement was 0.1 cm and the absolute error 0.05 cm).

Example
For each of the following measurements, find
   a) the smallest unit of measurement,
   b) the absolute error,
   c) the upper and lower limits of the true measurement.
      1. 5 cm

      2. 30.6 seconds

Solution
1. a) The smallest unit of measurement is 1 cm.
    b) The absolute error is half of the smallest unit, $= 0.5$ cm.
    c) The upper limit is 5.5 cm.
       The lower limit is 4.5 cm.

2. a) The smallest unit of measurement is 0.1 s
    b) The absolute error is half of 0.1 s $= 0.05$ s
    c) The upper limit is 30.65 s
       The lower limit is 30.55 s.

EXERCISE B

For each of the following measurements give
   a) the smallest unit of measurement,
   b) the absolute error,

c) the upper and lower limits of the true measurement.
   1. 7 cm   2. 186 m
   3. 17 hours   4. 9.2 kg
   5. 12.8 g   6. 2.683 litres
   7. 1.21 miles   8. 16.5 seconds

Relative error

The absolute error in itself is useful, but it is often more useful to find the **relative error.** This is found by considering the absolute error in relation to the measurement itself.

$$\text{Relative error} = \frac{\text{absolute error}}{\text{measurement}}$$

Example
Find the relative error in giving a length as 7.5 cm

Solution

$$\text{absolute error} = 0.05 \text{ cm}$$

$$\text{relative error} = \frac{0.05}{7.5}$$

$$= \frac{5}{750}$$

$$= \frac{1}{150}$$

This is often expressed as a percentage.

$$\text{percentage error} = \frac{1}{150} \times 100\%$$

$$= \frac{100}{150}\%$$

$$= \frac{2}{3}\%$$

$$= 0.67\%$$

Notice that whereas the absolute error is a quantity of the same kind as the measurement itself, the relative and percentage errors are numbers only.

In calculating absolute, relative and percentage errors, we always calculate the maximum possible errors and so in practice, the errors are likely to be less than those calculated.

EXERCISE C

Calculate for each of the following
measurements a) the absolute error
b) the relative error.

1. 150 m
2. 35 kg
3. 1.5 litres
4. 2.3 seconds

Calculate for each of the following
measurements a) the relative error
b) the percentage error
(to 2 significant figures)

5. 7 cm
6. 3.5 kg
7. 17 days
8. 25.0 g
9. 12.50 m
10. 150.00 litres

The sum and difference of measurements

Example 1
What are the upper and lower limits to the
sum of 6.4 cm and 1.2 cm (each given to 2
significant figures).

Solution

6.4 cm lies within the range 6.35 cm
and 6.45 cm

and, 1.2 cm lies within the range 1.15 cm
and 1.25 cm

Hence, the maximum sum is $6.45 + 1.25$
$$= 7.70 \text{ cm}$$

and, the minimum sum is $6.35 + 1.15$
$$= 7.50 \text{ cm}$$

Note: The apparent sum $(6.4 + 1.2)$ cm
= 7.6 cm, has a (maximum) error of
0.10 cm, this being the sum of the
absolute errors of the original
measurements, each of which was
0.05 cm.

Example 2
Find the limits within which lies the difference
of 6.5 kg and 3.25 kg.

Solution

6.5 kg lies within the range 6.45 kg
and 6.55 kg

and, 3.25 kg lies within the range 3.245 kg
and 3.255 kg

The maximum difference is $6.55 - 3.245$ kg
$$= 3.305 \text{ kg}$$

The minimum difference is $6.45 - 3.255$ kg
$$= 3.195 \text{ kg}$$

Note: The apparent difference $6.5 - 3.25$ kg
= 3.25 kg, has a (maximum) absolute
error of $(3.305 - 3.25)$ kg or
$(3.25 - 3.195)$ kg, both being equal to
0.055 which is the sum of the absolute
errors of the original measurements.

EXERCISE D

1. Find the upper and lower limits of the
true sum of the measurements given.
a) 6.5 cm and 8.2 cm
b) 15 g and 17 g
c) 17 m and 23 m
d) 16 litres and 23 litres

2. Without actually calculating the sum
write out the absolute errors in adding these
measurements:
a) 320 m and 47 m
b) 32 kg and 14 kg
c) 17.5 g and 5.3 g
d) 4.32 cm and 3.18 cm

3. What are the upper and lower limits of the
true differences between the following
measurements?
a) 19 cm and 5 cm
b) 13 kg and 8 kg
c) 13.5 m and 6.2 m
d) 17.6 litres and 14.0 litres

4. Without working out the differences write
down the absolute error in subtracting these
measurements.
a) 35 m and 17 m
b) 230 ml and 155 ml
c) 17.5 cm and 13 cm
d) 45 g and 24.5 g

The product and quotient of measurements

Example 1
Within what limits does the following product lie? $16 \text{ cm} \times 11 \text{ cm}$

Solution
16 cm lies within the range 15.5 cm and 16.5 cm

and

11 cm lies within the range 10.5 cm and 11.5 cm

$$\text{The maximum product} = 16.5 \times 11.5 \text{ cm}^2$$
$$= 189.75 \text{ cm}^2$$
$$\text{The minimum product} = 15.5 \times 10.5 \text{ cm}^2$$
$$= 162.75 \text{ cm}^2$$

Thus the true product lies between 189.75 cm² and 162.75 cm²

The apparent product is
$$16 \times 11 \text{ cm}^2 = 176 \text{ cm}^2$$
$$\text{and, } (189.75 - 176) \text{ cm}^2 = 13.75 \text{ cm}^2$$
$$(176 - 162.75) \text{ cm}^2 = 13.25 \text{ cm}^2$$

so that the (maximum) absolute error is
$$13.75 \text{ cm}^2$$

Example 2
Within what limits does this quotient lie?
$$195 \div 13$$

Solution
195 lies within the range 194.5 and 195.5

and, 13 lies within the range 12.5 and 13.5

The maximum quotient is
$$\frac{195.5}{12.5} = 15.64$$

and, the minimum quotient is
$$\frac{194.5}{13.5} = 14.41$$

Thus, the true quotient lies between 15.64 and 14.41

The apparent quotient
$$= \frac{195}{13} = 15$$

so, the absolute error is
$$(15.64 - 15) = 0.64$$

or
$$(15 - 14.41) = 0.59$$

i.e. the (maximum) absolute error is 0.64

EXERCISE E

Find the upper limits of the following expressions.

1. 32×15 2. $304 \div 16$ 3. $\dfrac{42 \times 13}{17}$

4. 14.5×11.3 5. $201.5 \div 19.3$ 6. $\dfrac{16.3 \times 13.4}{23.7}$

Calculate the (maximum) absolute error of these expressions.

7. 16×13 8. $210 \div 23$ (figures given correct to 2 significant figures)

9. $436 \div 19$ 10. $\dfrac{420 \times 32}{120}$ (figures given correct to 2 significant figures)

3 SAMPLING

Have you ever sampled your mother's baking? Presumably you did not eat all of a batch of cakes to see what the baking was like. You tasted one cake and assumed that they were all of the same quality. But what if the one you tasted was too near the heat in the oven and was burned? Was this typical of all the cakes or was it an exception? This is the kind of thing we must consider when we take a sample of something.

Sampling things is part and parcel of everyday life. A farmer will run his hand through a sack of grain and pick out a handful to examine. A grocer may give a woman a small piece of cheese to taste so that she can decide whether to buy some or not. Often manufacturers of breakfast cereals, soap powders, etc, will distribute small sample boxes of their products to housewives so that they can try them.

If you stop to think about it, you will find that almost all the facts we know have been discovered by sampling. We know what rivers are like because we have seen a few, not all of them. We know what poetry is like by reading some poems. Hit Parade ratings of popular songs are obtained by questioning a sample of people, or finding out how many records have been sold in selected record shops, not in all the record shops.

Sampling is of great importance to manufacturers. They must be constantly on the alert to ensure that their products are of a high standard. In the production of some goods, every article can be tested e.g. every

T.V. set completed in a factory can be switched on to see that it is working. But how impractical it would be to test every ball bearing made in a factory to see that it was of standard size.

Many tests are by their nature destructive of the article e.g. testing the length of life of a light bulb or eating a cake to see if it is of good quality. If every single article was tested in this way, there would be nothing left to sell.

Manufacturers take samples of each batch of articles made to check the quality of the product.

In all the situations mentioned above, we assume that the sample gives us information about the whole collection or 'population'. We must recognise that whenever we obtain information from a sample and assume that this information holds good for the whole 'population', we run the risk of making a mistake.

The **population** is the general term given to the aggregate of whatever we are considering. It only means a human population when people are being considered.

A numerical characteristic of a population such as its mean or standard deviation is called a **parameter**. A quantity calculated from a sample is called a **statistic**.

Since parameters of a given population are based upon **all** its variates, they are fixed for that population. On the other hand, since statistics are based upon only a **part** of the population, they usually vary from sample to sample.

Types of sample

1. Random sample

This is a sample in which each member of the population has an equal chance of being chosen. For example, a number of even-sized marbles are in a box, some red, some green. The box is given a good shake and five marbles are picked out without looking. A random sample of five marbles has been chosen, each marble in the box having had an equal chance of being picked.

In a card game, the cards are shuffled and dealt so that each player gets a random sample.

It may seem a simple matter to obtain a random sample but in practical situations, particularly where humans populations are concerned, it is far from simple.

Even in the card game mentioned above, what if the person shuffling and dealing the cards is a card-sharp? The cards dealt to each player would certainly not be random samples, they would be distinctly biased.

How to choose a random sample
If the population we are considering is small, we can allot a numbered card or disc to each member and, by mixing the cards thoroughly, pick out the required sample. Each member whose card is picked is in the sample.

This method, however, is not practical where the population is large. Here we can use tables of **random numbers.**

Random numbers
Sets of random numbers may be obtained from books of tables of random numbers (see Table 2 of Appendix). Or if a computer is readily available, it can be used to produce sets of random numbers.

These random numbers may be
single digits   e.g. 3 2 4 9 8 3 2 5
two-digit numbers   e.g. 13 51 07 88
three-digit numbers   e.g. 139 004 612
   etc.
A set of single-digit random numbers may be used to give two-digit, three-digit numbers etc. simply by splitting the row of digits into groups of two or three etc.
   e.g. 3 2 4 9 8 6 4 0 5 3 9 7 2 3 1
two-digit nos. 32 49 86 40 53 97 23
three-digit nos. 324 986 405 397 231

Use of random numbers to pick a sample
There is a population of 250 scores out of which a sample of 20 is to be chosen using random numbers. So, using a set of random numbers, we require three-digit numbers 250 **or less** i.e. any number above 250 is no use.

Using the set of numbers below
183 521 906 518 732 190 354 991 035 784 etc.
Any number above 250 is discarded and we go through the list until we have the required 20 suitable numbers.

The first suitable numbers from the above are 183 190 035.

The members of the population which have been given these numbers are those to be used in the sample.

2. Stratified random sample

This is a sample, made up of random samples from each section or stratum of a population, where the size of each of the random samples is proportional to the size of that section of the population.

3. Quota sample

This is the kind of sample often used in marked research surveys. Each interviewer is given definite instructions about the section of the public he is to question, but the final choice of actual persons is left to his own convenience. He might be told, for instance, to question 30 people, who are to include 3 men over 60 years of age; 5 men between 40 and 60; 6 men between 20 and 40; 6 men or boys under 20; 3 young schoolboys; 4 men dressed like business men; 3 men dressed in working clothes—overalls etc.

It is up to the interviewer which individual persons he questions. There will probably be a fairly large number of interviewers in various parts of the country conducting their surveys on the same day. The results of these are sent into a central office and collated there.

4. Systematic sample

To obtain a systematic sample, every nth member of the population is taken.

The value of "n" depends on the size of sample wanted and on the size of the population from which the sample is to be drawn. For example, if there were 2000 names on the electoral roll of a certain town and a sample of size 200 was to be chosen, then every tenth name on the roll would be selected. If, however, a sample of 100 was to be chosen, then every twentieth name would be picked.

5. Convenience sample

This is a sample comprising a part of the population which is "conveniently" close to hand and readily available.

Example
A certain secondary school has a roll of 900, made up as shown below.

Year	1	2	3	4	5	6
Number of pupils	192	183	210	175	92	48

We wish to pick a sample of 90 from the school, by
  a) random sample,
  b) stratified random sample.

Solution
a) Random sample—
i) Each pupil can be allotted a numbered card or disc. These are shaken in a box and 90 are picked out at random. The pupils with these numbers are the ones to be included in the sample.
(ii) Random numbers may be used to select the pupils.

b) Stratified random sample—
To obtain a stratified random sample of 90 from the school population, the number chosen from each year group must be in proportion to the numbers in each year group relative to the total roll.

Year	Fraction of pupils	Number of pupils
1	$\dfrac{192}{900}$	$\dfrac{192}{900} \times 90 = 19.2 = 19$
2	$\dfrac{183}{900}$	$\dfrac{183}{900} \times 90 = 18.3 = 18$
3	$\dfrac{210}{900}$	$\dfrac{210}{900} \times 90 = 21 \quad = 21$
4	$\dfrac{175}{900}$	$\dfrac{175}{900} \times 90 = 17.5 = 18$
5	$\dfrac{92}{900}$	$\dfrac{92}{900} \times 90 = 9.2 = 9$
6	$\dfrac{48}{900}$	$\dfrac{48}{900} \times 90 = 4.8 = 5$
		Total $= 90$

Note: Check that the "rounded off" figures give the correct sample size. If they do not, then a slight adjustment must be made to a suitable figure.

Once the number of pupils from each year group is calculated, then this set of pupils may be selected at random (see (a) above).

EXERCISE A

1. A statistics class want to choose a sample of 100 from a school where the numbers of pupils in each year are shown below.

Year	1	2	3	4	5	6
No. of pupils	290	285	310	175	92	48

  a) Explain how this sample could be obtained by picking a random sample.

b) If a stratified random sample is chosen, explain how this could be done and how many pupils from each year group are to be chosen for the sample.

2. In a small village, the population is divided by age groups as shown in the table.

Age (years)	0–4	5–15	16–45	46–65	65+
No. of people	15	41	50	70	14

It is proposed to choose a stratified random sample of 40 from the village. Explain how this should be done and calculate how many people should be picked from each age group.

3. In a primary school, the number of pupils in each class is shown below.

Class	1	2	3	4	5	6	7
No. of pupils	27	29	30	30	33	40	36

The Headmaster wants to pick a stratified random sample of 50 from the pupils. Explain how you would calculate the number of pupils to chosen from each class and give the numbers. Explain also how you would then pick the actual individual pupils in each class.

4. A factory making children's shoes completes 300 pairs in a day's work. The pairs are numbered from 1 to 300. It is wished to choose a sample of 10 pairs for testing purposes. Use the random numbers given below to show how this sample could be obtained.

2 9 5 2 9 7 9 2 7 9 7 9 7 0 0 2 8 1 2 6 4 1
6 7 2 7 6 2 7 2 0 3 5 9 1 1 6 1 1 1 3 2 7 0
6 1 0 7 3 5 6 3 5 3 5 6 3 1 7 0 8 2 1 4 6 4

5. Another factory, making crash helmets, completes 90 in a day's work. Use the random numbers from the last question to show how a sample of 5 helmets could be picked for testing.

6. A factory has 500 employees, each one having a "works number". For the purposes of a survey it is wished to pick a sample of 25 from the work-force.
Explain
a) how a systematic sample of 25 could be chosen,
b) how a random sample, using random numbers could be chosen,
c) how a random sample could be chosen, without the use of random numbers.

Random numbers:
3 5 6 3 6 0 0 8 7 0 0 2 5 9 1 1 5 3 5 6 1
0 8 9 8 1 2 6 6 1 1 1 2 0 5 6 0 2 3 7 0 4
1 6 7 2 9 5 2 9 7 9 2 2 7 6 2 6 1 0 7 9 0
2 5 4 3 1 4 7 2 0 3 7 9 7 9 9 7 9 2 2 7 6
2 5 2 0 9 0 7 3 2 7 6 1 4 3 0 2 7 1 3 3 2
6 2 1 8 2 0 7 0 1 1 5 9 6 5 3 5 7 0 1 6 7
6 1 4 0 6 5 1 6 7 1 9 2 4 1 2 6 2 7 8 3 1

7. Another small factory has 80 employees with numbers from 1 to 80. It is wished to select a sample of 10 employees for a survey.
Explain
a) how a systematic sample could be obtained,
b) how a random sample could be obtained using random numbers (use the random numbers from the last question).

26

FREQUENCY DISTRIBUTIONS

Variables

When we collect information to be used statistically we want to find out about a particular characteristic of a group of people or objects.

For example, we might want to know the heights of the boys in a class, the number of marks scored in an examination by the pupils of a class or the number of tomatoes on each plant in a certain greenhouse. The particular characteristic in which we are interested, is called the **variable** This variable 'varies' or changes from one member of the group to another.

Variables are of two main types—**qualitative** and **quantitative**. A qualitative variable is one which **describes** a characteristic e.g. the size of a person's feet—small or large. A quantitative variable is one which can be given a numerical value. Quantitative variables are of two distinct types—**discrete** or **continuous**.

A **discrete** variable is one which can only have certain definite values, very often whole numbers, e.g. the number of tomatoes on each plant in a greenhouse must be a definite value, in this case a whole number. There might be 2 tomatoes on one plant and 6 on another etc. but there cannot be 2.683 tomatoes on any plant. Some values of discrete variables are not in whole numbers. In shoe sizes there are the 'half' sizes. As a girl's foot grows she may take a size 1, $1\frac{1}{2}$ 2, $2\frac{1}{2}$ but no matter the actual length of the foot, she cannot buy a size 1.75. Shoes can only be bought in certain, distinct sizes.

The various values of a discrete variable can usually be obtained by **counting.** The number of tomatoes per plant can be counted. The number of girls who take a size 2 shoe can be counted.

Let us consider, now, the actual length of a girl's foot as it grows. At age 10 it measured 20 cm and at age 12 it was 25 cm. There is **no length** between 20 cm and 25 cm that the foot did not measure at some time. At one point in time it measured 21.25 cm, at another time it was 23.6847 cm etc. The length of the foot, then, varied continuously —there were no 'gaps' in its length between 20 cm and 25 cm.

A **continuous** variable is one which can take up any value within a certain range. The different values of a continuous variable are usually obtained by some kind of **measurement.** A boy's height or weight varies continuously and may be found by measuring.

EXERCISE A

State whether the following variables are qualitative or quantitative (discrete or continuous).

1. A baby's weight over the first year of its life.

2. The marks of a class in an English examination.

3. The colour of Mary's eyes (blue).

4. The temperature of a person who is ill.

5. The colour of John's hair (red).

6. The batting scores of a cricket team.

7. The speed of an aeroplane during a flight.

8. The length of a boy's left foot from age 5 to age 10.

9. The shoe sizes of the boy in question 8.

10. The number of goals scored by a football player during a football season.

11. The volume of water used by a town throughout the year.

12. The style of a girl's piano playing (very expressive).

13. Air pressures recorded at a weather station.

Frequency tables (ungrouped data)

The following marks are those scored by 50 pupils in an arithmetic test, the maximum mark being 10.

9	7	5	3	7	8	7	8	6	6
3	5	7	5	8	6	5	6	5	9
3	4	1	6	5	2	3	4	7	8
4	2	7	7	4	6	6	4	10	7
6	5	6	4	5	4	4	5	7	6

Data presented like this as a collection of single facts (a raw score) is not very useful. If you had sat that test and scored the mark of 6, you would like to know how your mark compared with the rest—whether it was a good mark or a poor one. This data must be tabulated. One way of doing this is to write down, in order, the possible marks and count up how many pupils scored each mark. The counting is done by means of tally marks. One stroke is marked down for each time the score occurs, every fifth stroke being drawn diagonally across the previous four, making groups of 5. If the score on the list is crossed out as it is tallied, it helps to avoid confusion. The number of times each

Scores in Arithmetic Test (above)

Score	Tally	Frequency
0		0
1	1	1
2	11	2
3	1111	4
4	⊥⊦⊦1 111	8
5	⊥⊦⊦1 1111	9
6	⊥⊦⊦1 ⊥⊦⊦1	10
7	⊥⊦⊦1 1111	9
8	1111	4
9	11	2
10	1	1
		$\overline{50}$

score occurs is called the frequency of the score.

You must always check that the total frequency is the same as the number of scores. If it does not come to the same answer, then you must check the tallying.

This table that we have made is called a **Frequency Distribution Table,** since it shows the frequency with which the various marks occur.

An important thing to take note of here is the **range** of the marks—in this case the marks range from 1 to 10, i.e. there is a range of 9 marks.

Now that we have tabulated the data it is much clearer what value a score of 6 has. 6 is, in fact, the score which occurred most often and this 'fashionable' score is called the **mode.**

16 pupils out of 50 or 32% of the group had a score greater than 6, and 24 pupils out of 50 or 48% of the group had a score lower than 6.

It is often important to know how the frequencies of the various scores compare to one another and to the total frequency.

The **relative frequency** of a score is the frequency of this score compared to the total frequency. For example, in the above table, the relative frequency of the score 6 is 10 out of 50, or $\frac{10}{50}$, or 20%, or 0.2.

EXERCISE B

1. These are the heights of 40, second year girls, measured to the nearest 2 cm. Make a frequency table and then answer the questions below.

154	144	152	158	158	154	148	150	160	154
156	152	150	154	158	150	154	156	142	154
152	152	154	152	156	148	150	154	154	152
150	158	156	158	158	154	164	148	152	152

a) State whether the variable here is continuous or discrete.

b) What is the range of the heights?

c) What is the modal height?

d) What is the relative frequency of 154 cm?

e) How many girls are 148 cm or less?

f) What percentage of the girls are taller than 158 cm?

g) What percentage of the girls are at least 150 cm tall but not more than 156 cm?

2. List the heights of the pupils in your own class and make a frequency table. Find out what the mode and the range of heights are. Compare your results with those in question 1. Explain any differences or similarities.

3. A pupil collected the following data from a third year class in November 1971.

These are the number of children in the families of the pupils questioned.

1	6	3	2	5	9	6	2	3	6
7	3	2	5	4	3	2	6	5	6
4	8	10	5	6	1	10	6	3	

Make a frequency table.

a) Is the variable continuous or discrete?

b) What is the mode?

c) What kind of family is not taken into account in this table? Why not?

d) How many of the pupils in this class belong to families with more than 6 children?

e) What percentage of the pupils are in families with 3 or less children? (answer correct to 1st decimal place).

4. Make a list of the number of children in the families of the pupils in your own class and make a frequency table.

Compare your results with those in the last question. Point out any similarities or differences.

5. These are the results of the Scottish Football matches (First and Second Division) one Saturday in the 1968–69 season.

2—0	2—1
1—1	0—2
2—4	2—0
3—1	1—3
2—1	1—3
3—1	0—2
1—0	2—1
4—0	5—1
2—0	2—4

a) Form a frequency table of the number of goals scored by each team.

b) Is this a continuous or discrete variable?

c) What is the range of goals scored?

d) What is the mode?

e) What is the relative frequency of the mode?

f) What percentage of the teams scored 3 or more goals?

6. Repeat the last questions with your own data about recent football results.

Histograms (ungrouped data)

One of the best ways of representing a frequency distribution graphically is by means of a **histogram.**

This is the frequency table that we made of the marks in the arithmetic test earlier in the chapter.

Score	0	1	2	3	4	5	6	7	8	9	10
Frequency	0	1	2	4	8	9	10	9	4	2	1

Fig. 4.1 Arithmetic Test

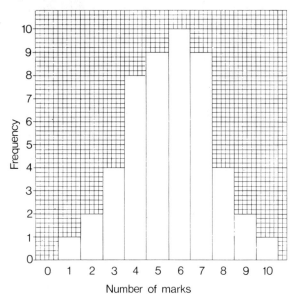

Frequency

Number of marks

The columns can be drawn without leaving any spaces between because there is a regular scale along the horizontal axis.

The values of the variable are always shown on the horizontal axis and the frequencies on the vertical axis.

EXERCISE C

1. The histogram below (Fig. 4.2) shows the number of points gained by the teams of the Second Division of the English football league at one stage of the 1968–69 season.

a) How many clubs are in this league?

b) What is the range of the points?

c) What is the mode number of points?

d) What is the relative frequency of 16 points?

e) How many clubs gained 22 or more points?

f) How many clubs gained less than 15 points?

As you see, a histogram is a column graph in which the areas of the rectangles are proportional to the frequencies.

Fig. 4.2 Points Scored By Teams of English Second Division

Frequency

Number of points

2. The following histogram illustrates the distribution of the shoe sizes of all boys aged 12–13 years in a school one January.

Fig. 4.3 Shoe Sizes of Boys Aged 12–13 Years

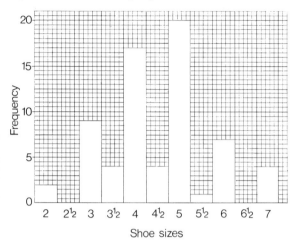

a) Do you notice anything unexpected about the distribution of the frequencies? Can you suggest any reason for this?

b) What is the total frequency of the distribution?

c) What is the modal shoe size?

d) What is the relative frequency of the modal size?

e) What is the relative frequency of size 6?

3. Draw histograms for the frequency distributions you tabulated for Exercise B.

4. The numbers of words in each line of a page of the book *Redgauntlet* by Sir Walter Scott are shown below. Form a frequency table and draw a histogram of the distribution.

11	10	6	8	12	1	9	5
7	8	8	8	8	9	10	8
8	8	3	7	13	5	10	9
7	9	6	8	6	8	11	
9	9	8	9	11	3	6	

a) How many lines are in the page?

b) What is the modal number of words per line?

c) What is the range of the distribution?

d) What is the relative frequency of 9 words?

e) How many lines have more than 9 words?

f) How many lines have less than 5 words?

g) How many lines have no more than 8 words and no less than 6?

5. Choose any page at random from a novel and count the number of words in each line. Construct a frequency table and draw the histogram for the distribution.

6. A class of first year pupils was given a speed and accuracy test consisting of ten sums—addition, subtraction, multiplication and division of number. The results obtained were as follows:

No. of sums correct	0	1	2	3	4	5	6	7	8	9	10
Frequency	0	1	2	6	11	5	2	2	1	2	0

Draw a histogram of the distribution.

a) How many pupils took the test?

b) What was the modal number of correct sums?

c) What was the range of the marks?

d) What percentage of the pupils had more than 5 sums correct?

Frequency polygons (ungrouped data)

Another way of representing a frequency distribution graphically is by means of a line graph called a **frequency polygon**. A frequency polygon is drawn by joining, with straight lines, the mid-points of the tops of the columns of the histogram. Figure 4.4 (a) shows a frequency polygon superimposed on

31

the histogram. Figure 4.4(b) shows the polygon by itself. To draw the polygon without drawing the histogram first, the points are plotted where the mid-points of the columns would have been.

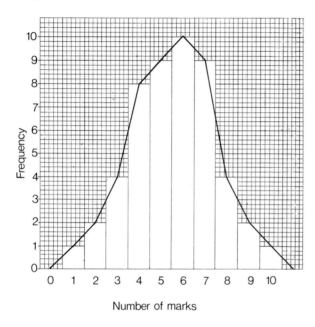

Fig. 4.4 (a) Arithmetic Test

Number of marks

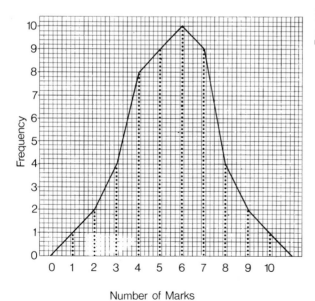

Fig. 4.4 (b) Arithmetic Test

Number of Marks

Since the area under the polygon must be equal to the area contained by the columns of the histogram, the polygon ought to be finished off by continuing the line at each end to the horizontal axis to where the next score would have been found.

Frequency polygons are very useful if we want to compare two distributions. It is difficult to draw two histograms on the same graph so that the distributions are clearly seen. However, it is very easy to draw two or more polygons on the same graph. The only drawback to this kind of comparison is that the total frequencies of the distributions must be the same, in order to give a fair comparison.

EXERCISE D

Fig. 4.5 Goals Scored By Teams of The English Football League One Saturday

Number of Goals

1. This polygon shows the number of goals scored by the teams of the four divisions of the English Football League one Saturday in the 1968–69 season.
   a) How many teams scored 1 goal?
   b) How many teams scored more than 2 goals?
   c) How many teams played that Saturday?

2. Draw frequency polygons on the histograms you drew for Exercise C.
3. Draw polygons by themselves for the distributions in Exercise C.
4. Fifty pupils were given two tests in successive weeks. The results were as follows:

Marks		0	1	2	3	4	5	6	7	8	9	10
Frequency	1st test	1	4	7	15	12	5	4	2	0	0	0
	2nd test	0	0	0	3	4	7	20	8	5	2	1

On the same graph draw frequency polygons for each distribution. (Use a solid line for the first test and a broken line for the second)
  a) What was the modal mark in each test?
  b) Comment on the way the marks are distributed. What reason might there be for it?
5. Measure the heights of 40 boys and 40 girls in the same year (to the nearest 2 cm). Form frequency tables and draw frequency polygons for each distribution on the same graph. Comment on the distributions.
ALTERNATIVELY—measure the heights of 40 boys (or girls) in one year and 40 boys (or girls) in another year and draw polygons for each distribution on the same graph.

Frequency tables (grouped data)

Sometimes the data we are considering has such a large range of scores that it is necessary to collect the scores into groups or classes. The **class interval** is the size of the group chosen. The class interval is decided by looking at the range of the scores and choosing the interval so that there are about 8 to 14 groups.

80	93	63	74	51	60	61	53	69	54
50	43	32	70	40	30	55	57	59	55
4	50	42	47	53	52	67	44	35	48
37	56	39	52	12	44	43	58	49	32
65	64	62	73	21	21	86	67	75	68

The figures at the bottom of the last column are a list of the marks of 50 boys in an examination. First we look through the list and pick out the highest and lowest scores. These are 93 and 4. A good way of grouping these marks is in tens—starting with the group 0–9, then 10–19, 20–29 etc. This will give us 10 groups. The scores are tallied in exactly the same way as for ungrouped data.

Mark	Tally	Frequency
0– 9	1	1
10–19	1	1
20–29	11	2
30–39	⊥⊥⊥⊤ 1	6
40–49	⊥⊥⊥⊤ 1111	9
50–59	⊥⊥⊥⊤ ⊥⊥⊥⊤ 1111	14
60–69	⊥⊥⊥⊤ ⊥⊥⊥⊤	10
70–79	1111	4
80–89	11	2
90–99	1	1

With grouped data we cannot pick out a single mark as the mode, but we can pick out the **modal class** or **group.** In this frequency distribution, the modal class is the group (50–59) marks.

One point that should be noted here is that a certain amount of accuracy is lost when data is grouped. For instance, from the table, we know that there are 6 scores between 30 and 39. Unless we have the original raw scores we do not know that the actual scores are 30, 32, 32, 35, 37 and 39.

Class boundaries

When selecting a grouping care must be taken to ensure that no gaps are left between the classes and also that the classes are not allowed to overlap. In the above example it would be useless to take classes 0–10, 10–20, 20–30, etc. as here we would not know whether to place a score of 10 in the first or second class. Neither of these difficulties will occur if we have a clear understanding of the nature of our data and of our class boundaries. In the above example the actual marks in the second class are 10–19, but we say that the boundaries enclosing the class are 9.5 and 19.5. If the first few classes are written in a row it helps to make this clearer.

Class	0–9	10–19	20–29	30–39	
Boundaries	9.5	19.5	29.5	39.5	

Thus 9.5 is the boundary separating the first and second classes i.e. 9.5 is the upper boundary of the first class and the lower boundary of the second class. We are in fact taking as our class boundary a measure which is half way between the largest measure we wish to include in the one class and the smallest we wish to include in the next. With most types of data this is quite straightforward but occasionally, especially when dealing with age, care must be taken e.g. the group of 13 year old children have ages ranging from 13 years exactly, up to but not including 14 years.

The mid-point of the class is now taken as the average of the upper and lower boundaries of the class. In the above example dealing with marks the mid-point of the second class is 14.5 and for the class of 13 year old children the mid-point of the class is 13.5 years. The mid-point of the class interval is important, as you will see later,

because it is often used to stand for the whole group.

EXERCISE E

1. Form a frequency table of the test marks of 40 pupils shown below, using class intervals of 5 marks starting at 1–5, 6–10 etc.

20	19	18	6	25	15	16	30	17	25
2	22	5	22	23	8	15	7	7	12
11	26	21	12	14	28	9	18	13	19
16	31	38	20	33	10	42	16	47	13

a) What is the mid-point of the 5th class?
b) What is the upper boundary of the 5th class?
c) What is the lower boundary of the 2nd class?
d) What is the modal class?
e) What is the relative frequency of the third class?
f) What percentage of the pupils scored more than 25 marks?
g) What percentage of the pupils scored more than 10 marks but less than 26?

2. Construct a frequency table for the examination marks of your own class in any subject and group the marks in suitable class intervals.

3. The table at the foot of the page shows the infant mortality rates (per 1000 live births) in Scotland for each quarter year, over several years. Construct a frequency table of the quarterly rates choosing suitable class intervals.

Year	1954	–55	–56	–57	–58	–59	–60	–61	–62	–63	–64
1st Quarter	38	36	31	32	32	33	29	30	30	29	27
2nd ,,	30	29	26	27	27	26	26	25	26	24	23
3rd ,,	24	26	27	27	25	25	24	23	24	21	22
4th ,,	31	31	29	29	27	29	27	26	26	28	23

a) What is the mid-point of the 1st class?
b) What is the upper boundary of the 3rd class?
c) What is the lower boundary of the 5th class?
d) What is the modal class?
e) What is the relative frequency of the 5th class?

4. The following figures were the number of rain days (days in which rain fell) recorded at selected stations all over the U.K. in one year:

260	234	209	241	243	268	185
241	253	228	199	158	236	209
173	187	191	207	199	199	172
178	157	188	148	183	160	152
197	171	265	226	235	226	233
180	182	211	202	185	206	209
182	188	164	156	160	250	220
187	195	215	177	191	225	164
209	206	194	205	195	198	177
175	180					

Construct a frequency table using class intervals of 10 starting at 140–149.

a) What is the mid-point of the 4th class?

b) What is the lower boundary of the 6th class?

c) What is the upper boundary of class 10?

d) What is the modal class?

e) What number of stations recorded less than 200 rain days?

f) What percentage of stations recorded at least 170 rain days and not more than 209?

Histograms (grouped data)

To represent graphically frequency distributions with grouped data, we again use histograms. However with grouped data, we must be very careful how we use the horizontal scale.

Below is the frequency distribution of marks that we tallied in the previous section.

Mark	0–9	10–19	20–29	30–39	40–49	50–59	60–69	70–79	80–89	90–99
Frequency	1	1	2	6	9	14	10	4	2	1

Fig. 4.6 Examination Marks

Marks

35

Take particular note of the scale used on the horizontal axis in Figure 4.6. On the actual scale, the class boundaries are used, and the groups are shown above. As you become used to drawing these histograms you should start to leave out the groups and simply show the boundaries.

EXERCISE F

1. The histogram shows the distribution of the rainfall (in cm) recorded at selected stations all over Britain, one year.

Fig. 4.7 Rainfall Recorded in Britain For a Year

a) How many stations were there?
b) What is the upper boundary of the 1st class?
c) What is the lower boundary of the 5th class?
d) What is the mid-point of the 3rd class?
e) What is the modal class?
f) What is the relative frequency of the modal class?
g) What percentage of the stations recorded a rainfall of more than 125 cm?
h) What percentage of the stations recorded a rainfall of 50 cm or less?

2. The distribution of the weights (in kg) of a number of boys is shown in the histogram Figure 4.8.
   a) How many boys were weighed?

Fig. 4.8 Weights of Boys (in Kg)

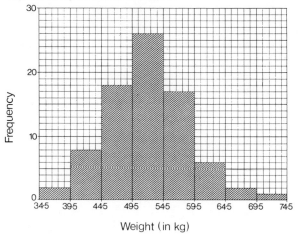

b) What is the upper boundary for the 3rd class?
c) What is the lower boundary for the 3rd class?
d) What weights are in the 3rd class?
e) What weights are in the 5th class?
f) What is the mid-point of the 4th class?
g) What is the modal class?
h) What is the relative frequency of the modal class?

3. Draw histograms for all the distributions you tabulated in Exercise E.

4.

Age	Frequency
0– 4	37
5– 9	47
10–14	19
15–19	9
20–24	5
25–29	4
30–34	11
35–39	13
40–44	8
45–49	11
50–54	20
55–59	23
60–64	33
65–69	30
70–74	30
75–79	28
80–84	28
85+	10

The table on the previous page shows the number of pedestrians killed in one year in Scotland in motor vehicle accidents, according to age in years.

Draw a histogram to illustrate this, without any further grouping.

Comment on the shape of the distribution and give reasons for it.

Frequency polygons (grouped data)

Frequency polygons are used to show frequency distributions with grouped data, as line graphs. To draw a polygon we join up with straight lines the mid-points of the tops of the columns of the histogram.

Example 1
The marks of some pupils in a test are shown below grouped in class intervals of 5.

You can simply draw this like any line graph, plotting the mid-points against the corresponding frequency—see Figure 4.9 (b).

Fig. 4.9 (a) Test Results

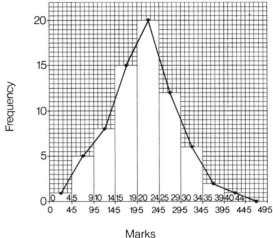

Mark	0–4	5–9	10–14	15–19	20–24	25–29	30–34	35–39	40–44	45–49
Frequency	1	5	8	15	20	12	6	2	1	0

Figure 4.9 (a) shows the polygon of this distribution drawn on top of the histogram.

Figure 4.9 (b) shows the polygon by itself.

If you study Figure 4.9 (a) you will see that the key points for the polygon are actually above the mid-points of the class intervals, and to draw the polygon on a graph by itself you need only plot the mid-points of the intervals against the corresponding frequencies.

The mid-points of the intervals are shown below.

Fig. 4.9 (b) Test Results

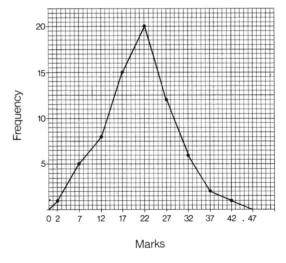

Mid-point of interval	2	7	12	17	22	27	32	37	42	47
Frequency	1	5	8	15	20	12	6	2	1	0

Example 2
Draw the frequency polygon for the following
distribution of examination marks.

Mark	0–9	10–19	20–29	30–39	40–49	50–59	60–69	70–79	80–89	90–99
Frequency	1	1	2	6	9	14	10	4	2	1

Solution
First find the mid-points of the class intervals

Mid-point	4.5	14.5	24.5	34.5	44.5	54.5	64.5	74.5	84.5	94.5
Frequency	1	1	2	6	9	14	10	4	2	1

Fig. 4.10 Examination Marks

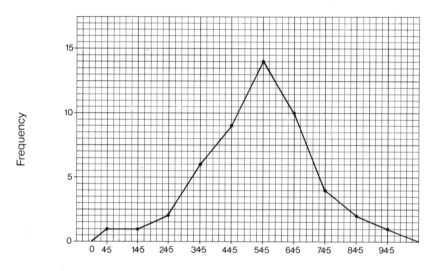

Marks

EXERCISE G

1. Draw frequency polygons superimposed
on the histograms you drew for Exercise F.

2. Draw polygons by themselves (adjusting
your scale accordingly) for the distributions
in Exercise F.

3. The heights of 50 second year boys were
measured to the nearest cm in October 1967.
In October 1968 the heights of 50 boys in
the second year were again measured (to the
nearest cm). These heights are shown below.
   Draw, on the same graph, frequency
polygons for each distribution.
   Comment on any differences you find.

Height (cm)		116–120	121–125	126–130	131–135	136–140	141–145	146–150	151–155	156–160	161–165	166–170	171–175
Freq.	1967	2	0	0	0	0	5	7	14	11	8	2	1
	1968	0	0	0	0	1	3	10	20	12	3	1	0

4. The number of words in each of the first hundred sentences of two books were counted. One book was 'Moby Dick' by Herman Melville (the story of the giant whale) first published in 1851. The other book was a modern, run of the mill, detective story first published some hundred years later.

The distributions for the number of words are shown below.

Draw polygons for the distributions on the same graph and comment on the way the distributions are dispersed.

What does the graph indicate to you about the 'readability' of the two books.

Moby Dick

No. of words	Frequency
1– 10	30
11– 20	23
21– 30	22
31– 40	15
41– 50	4
51– 60	0
61– 70	3
71– 80	1
81– 90	1
91–100	0
101–110	0
111–120	1

Detective Story

No. of words	Frequency
1–10	62
11–20	24
21–30	10
31–40	2
41–50	0
51–60	1
61–70	1

5. Choose 2 novels (*not* textbooks) and count the number of words in each of the first 100 sentences and repeat question 4.

6. Shown below are the number of hours of bright sunshine recorded in 52 selected stations all over Scotland one year, in the months of December and June.

Tabulate the data for each month grouping in suitable class intervals, and draw a frequency polygon for each month on the same graph.

Compare and contrast the two distributions.

December—No. of hours of bright sunshine

16	25	41	20	35	20	16	8	38	23
25	38	38	41	34	39	24	47	45	17
42	44	47	45	51	35	37	51	47	42
42	31	39	55	45	42	33	27	35	40
28	35	39	36	14	14	40	44	43	34
50	43								

June—No. of hours of bright sunshine

163	153	164	137	177	160	158	148
171	152	169	180	179	152	164	157
181	198	188	163	195	195	209	212
195	204	197	215	206	192	201	149
186	234	192	177	144	120	186	173
176	163	177	175	163	148	161	172
150	136	150	151				

Cumulative frequency

It is often useful to know how many boys have scored more than a certain mark in an examination or how many girls are less than a particular height. An easy way to answer this type of question is by making a **cumulative frequency** table. A frequency table is converted to a cumulative frequency table by adding each frequency to the total of its predecessors.

Example 1
The marks of 30 pupils in a test are shown below (a).

The cumulative frequency table for these marks is shown in (b).

(a)

Mark	Frequency
0	0
1	2
2	3
3	4
4	6
5	8
6	5
7	1
8	1
9	0
10	0

(b)

Mark		Cumulative Frequency	
	0		0
up to and including 1		(0+2)	2
„ „ „ „	2	(2+3)	5
„ „ „ „	3	(5+4)	9
„ „ „ „	4	(9+6)	15
„ „ „ „	5	(15+8)	23
„ „ „ „	6	(23+5)	28
„ „ „ „	7	(28+1)	29
„ „ „ „	8	(29+1)	30
„ „ „ „	9	(30+0)	30
„ „ „ „	10	(30+0)	30

From the cumulative frequency table we can see, for example, that

23 pupils scored 5 or less marks.
9 pupils scored less than 4 marks.

Example 2
In the next column (a) is the frequency table we constructed a few pages ago of the marks of 50 boys in an examination.

(b) shows the cumulative frequency table for the distribution.

(a)

Mark	Frequency
0– 9	1
10–19	1
20–29	2
30–39	6
40–49	9
50–59	14
60–69	10
70–79	4
80–89	2
90–99	1
	50

(b)

Mark		Cumulative Frequency	
up to and including	9		1
„ „ „ „	19	(1+1)	2
„ „ „ „	29	(2+2)	4
„ „ „ „	39	(4+6)	10
„ „ „ „	49	(10+9)	19
„ „ „ „	59	(19+14)	33
„ „ „ „	69	(33+10)	43
„ „ „ „	79	(43+4)	47
„ „ „ „	89	(47+2)	49
„ „ „ „	99	(49+1)	50

Instead of writing
'up to and including 9'
„ „ „ „ 19', etc.,
we can show the mark for the cumulative frequency as

0– 9
0–19
0–29 etc.

From the above cumulative frequency we can answer these questions readily.

1. How many boys scored less than 50 marks? Answer—19

2. How many boys scored more than 59 marks? Answer—17 (50–33)

Height (cm)	146	148	150	152	154	156	158	160	162	164	166	168
Frequency	1	0	1	3	3	4	3	5	6	1	2	1

EXERCISE H

1. At the top of the page are the heights of a class of 30 fourth year girls measured to the nearest 2 cm.
Construct a cumulative frequency table and answer the following questions
   a) How many of the girls are less than 150 cm tall?
   b) What percentage of the girls are at least 158 cm in height?
   c) How many girls are over 164 cm?

2. The frequency distribution for the times of some telephone calls was as follows.

Time (s)	Frequency
0– 20	0
21– 40	2
41– 60	4
61– 80	7
81–100	11
101–120	28
121–140	20
141–160	16
161–180	10
181–200	6
201–220	3

   a) Construct a cumulative frequency table for the distribution.
   b) What number of calls lasted a minute or less?
   c) What percentage of the calls lasted more than 2 minutes?

3. The frequency distribution below shows the weights of fourth year boys measured to the nearest kg.

Construct a cumulative frequency table.
   a) What was the total number of boys weighed?
   b) How many boys weighed less than 50 kg?
   c) How many boys weighed 60 kg or more?

4. In a major Golf Tournament, one September, the leading ten players handed in the following scores (4 rounds each).

70	69	68	69	66	68	71	71	71	69
70	69	71	74	75	68	70	74	72	68
69	70	72	68	70	73	69	67	75	73
66	68	67	68	68	70	70	69	65	71

   a) Construct a frequency table for the scores.
   b) Convert this to a cumulative frequency table.
   c) How many rounds of less than 70 were there?
   d) How many rounds of more than 71 were there?

Cumulative frequency curves

When we draw the graph of a cumulative frequency distribution we obtain a curve which has a characteristic shape. This curve is called a **cumulative frequency curve** (or **ogive**—from a term used in architecture for this shape of curve).
   Figure 4.11 shows the cumulative frequency curve for example 1 of the last section.

Weight (kg)	35–39	40–44	45–49	50–54	55–59	60–64	65–69	70–74	75–79	80–84
Frequency	1	2	3	8	8	4	1	0	1	1

Fig. 4.11 Test Results

Fig. 4.12 Examination Results

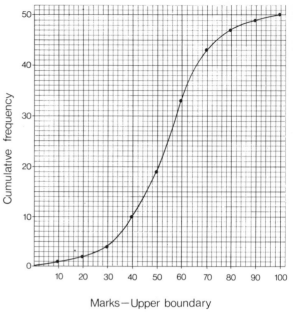

Mark	Cumulative Frequency
0	0
0– 1	2
0– 2	5
0– 3	9
0– 4	15
0– 5	23
0– 6	28
0– 7	29
0– 8	30
0– 9	30
0–10	30

To plot the points of the graph, the cumulative frequency is drawn against the upper boundary of the class

e.g.

2 is plotted against the mark 1.5

5 against 2.5

9 ,, 3.5

15 ,, 4.5

Note: The horizontal scale is marked off in tens but the points are plotted at 9.5, 19.5 etc.

As with any line graph, 2 or more curves may be drawn on the same graph, for purposes of comparison.

Figure 4.11 shows the cumulative frequency curve of Example 2 of the last section.

Mark	Cumulative Frequency
0– 9	1
0–19	2
0–29	4
0–39	10
0–49	19
0–59	33
0–69	43
0–79	47
0–89	49
0–99	50

To plot the points of this graph we plot the cumulative frequency against the upper boundary of the class.

e.g.

1 is plotted against the mark 9.5

2 against 19.5

4 ,, 29.5

10 ,, 39.5

EXERCISE I

Draw cumulative frequency curves for all cumulative frequency distributions you tabulated in Exercise H.

EXERCISE J Miscellaneous

1. State whether the following variables are qualitative or quantitative (discrete or continuous).

   a) The lengths of the holes on a golf course.

   b) The number of putts taken at the various holes by a player.

   c) The type of weather (sunny).

d) The length of time taken to play a round of golf.

2. The length of each trout in a box was measured, correct to the nearest centimetre, and the results are illustrated below in the form of a cumulative frequency curve.

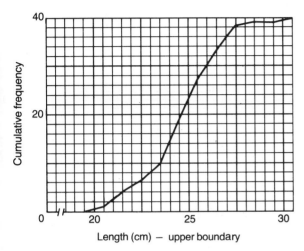

Fig. 4.13

a) Use the graph to draw up the cumulative frequency distribution.
b) Use your result above to obtain the frequency distribution of the lengths.
c) How many trout were less than 25 cm long?
d) How many trout were 27 cm or more in length?
e) What was the modal length?
f) What was the lower limit of the modal class?

3. At a busy airport the scheduled take-off times of aircraft over a three-hour period were as below.

0802	0810	0820	0825	0829	0835
0845	0853	0900	0909	0921	0924
0935	0948	1000	1010	1020	1029
1040	1043	1048	1055	1100	

a) List the intervals between the scheduled take-off times.
b) Construct a frequency tables for these times.

c) Illustrate the information by means of a histogram.
d) What is the modal interval between the scheduled take-off times?
e) How many of the intervals were less than the modal interval?

4. The weights of 40 fourth year pupils were measured to th nearest kilogramme and the results are given in the following table.

Weight (kg)	35–39	40–44	45–49	50–54	55–59	60–64	65–69	70–74
Frequency	1	6	9	10	6	4	3	1

The weights of the girls in the group are given in the following frequency polygon.

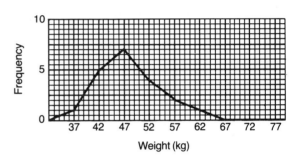

Fig. 4.14

a) Use the two sets of data to obtain the frequency distribution of the weights of the boys in the group.
b) What is the modal class for the weight of a boy?
c) What is the modal class for the weight of a girl?
d) What are the limits between which the weights of girls in the modal class must lie?
e) Draw up separate cumulative frequency distributions for the weights of boys and girls. Draw the two cumulative frequency curves on the same grid and comment on the result.

5. A histogram was drawn to illustrate the distribution of pupils within the various years of the school. The years and the heights of

43

the corresponding columns are given in the table below.

Year	1	2	3	4	5	6
Height (cm)	12.5	13.5	13	11.5	9.5	3

If the number of pupils in the sixth year was 36 find the number of pupils in each of the other years.

6. The lifetime of 100 electric bulbs was measured and the details are given below.

Lifetime (hours)	1000–1199	1200–1399	1400–1599	1600–1799
No. of bulbs	8	16	28	32

Lifetime (hours)	1800–1999	2000–2199	2200–2399
No. of bulbs	12	4	0

  a) What is the modal lifetime?
  b) Illustrate the information by means of a frequency polygon.
  c) Construct a cumulative frequency distribution from the above facts.
  d) How many of the bulbs had a lifetime which was less than that of the modal class?
  e) Construct the cumulative frequency curve.

7. The child population (to the nearest hundred) of an African country for a year in the 'sixties was as follows.

Age	Under 2	2–6	7–14
Child population	6200	9500	8800

Illustrate the information in the form of a histogram using axes as given below (note that the width of the columns will vary but the area of the column will have to correspond to the frequency).

Fig. 4.15

44

MEASURES OF CENTRAL TENDENCY

Data can be more easily understood when it is tabulated in an orderly fashion in a frequency distribution and then shown graphically. But one single value which characterises the group and which is most easily grasped by the mind, is the **average** of the group.

There are three 'averages' used in statistics:

      1. The Arithmetic Mean
      2. The Median
      3. The Mode

The arithmetic mean

This is the 'average' used in arithmetic but in Statistics we must be careful always to call it the mean.

To find the mean of set of scores, we simply add up the scores and divide by the number of scores.

Example
Find the mean of the following numbers.
1, 3, 5, 7, 9, 11, 13, 15.

Solution

$$\text{The mean} = \frac{\text{Sum of the scores}}{\text{number of scores}}$$

$$= \frac{1+3+5+7+9+11+13+15}{8}$$

$$= \frac{64}{8}$$

$$= 8$$

The formula we use to calculate the mean is,

$$\overline{X} = \frac{\sum X}{N}$$

where $\overline{X}$ = mean

    $\sum$ means sum of

    X represents the single scores

    N represents the number of scores.

The median

When a number of scores are arranged in **numerical order,** the median score is the 'middle' score having the same number of scores above it as below. When there is an odd number of scores this is easy as there is a 'middle' score but when there is an even number of scores there is no single 'middle' score and we define the median as the average of the two 'middle' scores.

Example 1
Find the median of the following golf scores.
    72, 68, 65, 70, 75, 79, 73

Solution
First arrange the scores in numerical order.

  65, 68, 70, |72,| 73, 75, 79,
                  median

  72 is the median since it has 3 scores above it and 3 below it.

Example 2
Find the median of these boys' heights
(measured to the nearest cm)

   148,   161,   167,   157,   162,   154.

Solution
First arrange in numerical order.

   148,   154,   157, | 161,   162,   167.
              median

   This time there is an even number of
scores and the median is the average of 157
and 161, i.e. the median is

$$\frac{157 + 161}{2} = 159.$$

The mode

The mode, as we saw in the last chapter, is
the score which occurs most frequently.

EXERCISE A

1. Find the mean, median and mode of the
following sets of data.

   a) 15, 11, 6, 3, 2, 14, 13, 7, 10, 11, 9.
   b) 24, 21, 20, 25, 21, 27.
   c) 5, 2, 2, 1, 3, 3, 3, 4, 4, 3, 3, 4, 2, 3, 2, 4.

2. In successive rounds a golfer took the
following numbers of strokes
         90, 69, 70, 70, 73, 71, 80.
Which of the three averages—mean, median
or mode—would he prefer to call his
'average' score?

3. The heights (to the nearest cm) of some
boys are given below.
162, 159, 162, 167, 165, 158, 166, 164, 158, 160.

What is *a)* the median height,
         b) the modal height,
         c) the mean height?

4. Make a list of 5 numbers, 4 of which are
smaller than the mean.

46

5. Make a list of 6 numbers, 5 of which are
greater than the mean.

6. If you were handed a box of used pencils
of varying lengths, what would be the easiest
average—mean, median or mode—to find,
and what would be the hardest?

7. In his savings account a boy had the
following balances at the ends of 6 successive
months. What was his mean balance and his
median balance?
      £6.50, £9.75, £4.75, £5.90, £8.40, £5.50.

Averages from frequency distributions

In the previous exercise, the mean, median
and mode were being obtained from raw
scores, but we often want to know these
averages from data which can be tabulated
into, or is already in the form of, a frequency
distribution.

The mean from a frequency distribution (ungrouped data)

Example
Calculate the mean of the following test
marks of 50 pupils (possible mark being 10)

   a) without tabulating them,
   b) by constructing a frequency table.

1,	5,	5,	3,	4,	6,	5,	6,	4,	7
·7,	6,	6,	7,	4,	7,	2,	3,	5,	6
8,	4,	2,	6,	7,	4,	7,	5,	6,	4
7,	9,	7,	3,	5,	8,	3,	5,	4,	8
4,	6,	6,	10,	7,	6,	5,	9,	8,	5

Solution

a)   The mean $\bar{X} = \frac{\sum X}{N}$

$$= \frac{277}{50}$$

$$= 5.54$$

b)

Score	Tally	Frequency
0		0
1	1	1
2	11	2
3	1111	4
4	1111 111	8
5	1111 1111	9
6	1111 1111	10
7	1111 1111	9
8	1111	4
9	11	2
10	1	1
		$\sum f = 50$

$$\overline{X} = \frac{(0 \times 0) + (1 \times 1) + (2 \times 2) + (4 \times 3)}{50}$$
$$\qquad \frac{+ (8 \times 4) + (9 \times 5) + (10 \times 6) + (9 \times 7)}{}$$
$$\qquad \frac{+ (4 \times 8) + (2 \times 9) + (1 \times 10)}{}$$

$$= \frac{277}{50}$$

$$= 5.54$$

To obtain 277 what we are doing is multiplying the frequency of each score by the score and adding the results.

An easier way of doing this is to add another column to the frequency table in which we multiply the frequency (f) by the score (X)

Score (X)	Frequency (f)	$f \times X$	
0	0	(0×0)	0
1	1	(1×1)	1
2	2	(2×2)	4
3	4	(4×3)	12
4	8	(8×4)	32
5	9	(9×5)	45
6	10	(10×6)	60
7	9	(9×7)	63
8	4	(4×8)	32
9	2	(2×9)	18
10	1	(1×10)	10
	$\sum f = 50$	$\sum fX = 277$	

$$\overline{X} = \frac{\sum fX}{\sum f} \quad (\sum fX = \text{sum of frequency} \times \text{score}$$
$$\qquad\qquad \sum f = \text{sum of frequencies})$$

$$= \frac{277}{50}$$

$$= 5.54$$

EXERCISE B

1. This frequency table shows the number of goals scored by the teams of the First and Second Division of the English Football League on Saturday 5th October 1968.

Number of goals	0	1	2	3	4	5
Frequency	15	16	4	6	2	1

Calculate the mean number of goals scored.

2. Construct a frequency table from the goals scored by the English First and Second Division teams last Saturday and calculate the mean number of goals.

3. The following table shows the number of children per family in the families of the pupils in a first year class.

Number of children	1	2	3	4	5	6	7	8	9
Frequency	2	4	10	6	6	3	2	1	0

Calculate the mean number of children per family.

4. Construct a frequency table of the number of children in the families of the pupils in your class and calculate the mean. (If you have already constructed such a table for question 4 in Exercise B of Chapter 3, then use it again.)

5. The following numbers are the number of words in each line of a page chosen from the novel *Westward Ho!* by Charles Kingsley.

7	7	7	10	13	8	10	6	9	12
11	1	12	13	2	11	10	10	13	9
12	1	10	8	12	12	12	12	3	7
10	12	9	10	12	11	12	9	2	11
8	13								

Construct a frequency table and then calculate the mean number of words per line.

6. Repeat the last question by choosing any novel and counting the number of words in each line of a page chosen at random.

The mean from a frequency distribution (grouped data)

Example

Calculate the mean of the following test marks (possible mark 50)
  a) without grouping
  b) by grouping in class intervals of 5.

4	23	35	27	32	15	29	19	25	18
36	11	33	12	6	30	20	24	27	25
16	40	21	35	20	17	9	30	14	29
31	26	23	25	21	32	23	18	26	11
28	43	24	43	24	10	37	22	34	23

Solution

a)

$$\bar{X} = \frac{\sum X}{N}$$

$$= \frac{1206}{50}$$

$$= 24.12$$

b)

Mark	Tally	Frequency
1–5	1	1
6–10	111	3
11–15	卌	5
16–20	卌 11	7
21–25	卌 卌 111	13
26–30	卌 1111	9
31–35	卌 11	7
36–40	111	3
41–45	11	2
46–50		0
		$\sum f = 50$

When we have a frequency distribution with grouped data, we use the mid-point of the interval to stand for the group, and as before, we add a frequency × score column to the table.

To calculate the mean, we again use the formula

$$\bar{X} = \frac{\sum fX}{\sum f} \quad \text{(This time X is the mid-point of the interval.)}$$

Mark	Mid-point of interval (X)	Freq. (f)	fX
1–5	3	1	3
6–10	8	3	24
11–15	13	5	65
16–20	18	7	126
21–25	23	13	299
26–30	28	9	252
31–35	33	7	231
36–40	38	3	114
41–45	43	2	86
46–50	48	0	0
		$\sum f = 50$	$\sum fX = 1200$

$$\bar{X} = \frac{\sum fX}{\sum f}$$

$$= \frac{1200}{50}$$

$$= 24.0$$

Note: When the data is grouped the mean is slightly different from that obtained when we use the raw scores. This is because of the slight loss of accuracy any time data is grouped. The more correct mean is that obtained from the raw scores.

EXERCISE C

1. This table shows the marks gained by 50 pupils in an arithmetic examination. Calculate the mean mark.

Mark	20–29	30–39	40–49	50–59	60–69	70–79	80–89
Frequency	1	4	8	15	16	5	1

2. These are the number of deaths due to railway accidents in Scotland in 1964, according to the age of the deceased person (in years):

Age	5–9	10–14	15–19	20–24	25–29	30–34	35–39
Frequency	1	1	3	2	1	1	1

Age	40–44	45–49	50–54	55–59	60–64	65–70
Frequency	1	1	1	5	2	1

Calculate the mean age of death.

3. The age distribution of workers in a factory was as follows. Calculate the mean age.

Age (years)	Frequency
16–20	2
21–25	10
26–30	12
31–35	17
36–40	15
41–45	14
46–50	9
51–55	8
56–60	8
61–65	5

4. These figures show the number of hours of bright sunshine recorded in 52 places in Britain one year for the month of October.

50	81	75	90	62	80	92	75	77	74
86	99	75	63	67	73	95	75	71	74
93	94	76	68	79	81	105	83	81	75
71	82	79	76	69	85	75	80	81	47
99	68	71	74	74	85	65	71	74	80
69	68								

Construct a frequency table using appropriate class intervals and calculate the mean number of hours of bright sunshine.

5. The times taken by rats to pass through a maze are recorded below. Calculate the mean time taken (to one decimal place).

Time (sec.)	10–14	15–19	20–24	25–29	30–34	35–39
Frequency	3	11	19	22	6	2

6. A manufacturer of cornflour wanted to check the accuracy of the machines used to fill the packets with cornflour. A sample of 100 packets was selected at random from the conveyor belt and these packets were weighed carefully. The results are shown in the table below. Calculate the mean weight of the packets.

Weight (grammes)	Frequency
450–459	1
460–469	7
470–479	40
480–489	44
490–499	6
500–509	2

An alternative method of calculating the mean using an assumed mean

This method of calculating the mean cuts down considerably on the amount of multiplication but care must be taken in using it.

In this method, we choose an estimated or assumed mean and work from it as shown below. We find the deviation of each score from the assumed mean and then find the mean of these deviations and apply this to the assumed mean to obtain the true mean.

Example 1
Here are the examination marks of 8 pupils:
    56, 58, 59, 60, 62, 63, 64, 65.
Find the mean mark, using an assumed mean.

Solution

Mark	Assumed Mean	Deviation from Assumed Mean (d)
56		−4
58		−2
59		−1
		−7
60	60	0
62		+2
63		+3
64		+4
65		+5
		+14

$$\sum d = -7 + 14$$
$$= +7$$

$$\text{Mean} = \text{Assumed Mean} + \frac{\sum d}{N}$$

$$= 60 + \frac{7}{8}$$

$$= 60 + 0.875$$

$$= 60.875$$

The approximate mean mark = 60.9

Example 2

The table below shows the heights of 26 pupils measured to the nearest centimetre. Find the mean height using an assumed mean of 152 cm.

Solution

(When there is a frequency for the scores, then a frequency × deviation column must be added to the table. This time we find the mean of the f × d column.)

$$\text{Mean} = \text{assumed mean} + \frac{\sum fd}{\sum f}$$

$$= 152 + \frac{9}{26}$$

$$= 152 + 0.35$$

$$= 152.35$$

The approximate mean height = 152.4 cm.

Grouped data

With grouped data we use the mid-point of the interval to stand for the whole group and we can further simplify the calculation by using as the unit of deviation from the assumed mean the band width. Here then d will measure band widths and when calculating the mean we must multiply $\sum fd$ by the band width to get the total deviate.

Height (cm)	Frequency (f)	Assumed Mean	Deviation from Assumed Mean (d)	f × d
148	1		−4	−4
149	1		−3	−3
150	2		−2	−4
151	3		−1	−3
				−14
152	6	152	0	0
153	7		+1	+7
154	3		+2	+6
155	2		+3	+6
156	1		+4	+4
	$\sum f = 26$			+23

$$\sum fd = -14 + 23$$
$$= +9$$

Example 3

The rateable values of eighty houses in one street were found and the results shown below. Calculate the mean rateable value using an assumed mean.

Rateable value (£)	50–59	60–69	70–79	80–89	90–99	100–109	110–119
Frequency	2	10	16	24	16	11	1

Solution

Rateable Value (£)	Frequency (f)	Assumed Mean	d	fd
50– 59	2		−3	−6
60– 69	10		−2	−20
70– 79	16		−1	−16
				−42
80– 89	24	84.5	0	0
90– 99	16		+1	+16
100–109	11		+2	+22
110–119	1		+3	+ 3
				+41
	Σf = 80			Σfd = −42+41 = −1

$$\text{Mean} = \text{Assumed mean} + \frac{\Sigma\,fd \times \text{band width}}{\Sigma\,f} = 84.5 + \frac{-1 \times 10}{80} \quad \text{(band width here is 10)}$$

$$= 84.5 - 0.125 = 84.375$$

The approximate mean rateable value = £84.38

EXERCISE D

1. Using an assumed mean of 100, calculate the mean of the following numbers.
   105, 108, 100, 94, 96, 97, 89, 111.

2. Using an assumed mean height of 148 cms, calculate the mean of these heights (measured in cms)—
   140, 142, 151, 150, 148, 149, 146, 145.

3. These are the actual weights (to the nearest tenth of a gramme) of 12 small bags of sweets. Calculate the mean weight, using an assumed mean. Weights—

   116.2 116.3 115.9 116 116.1 115.8

   116 116.2 116.3 116.1 116.2 115.9

4. The number of ships arriving in Glasgow from overseas, each month in a recent year was as follows:

   111 107 133 114 140 132 125 135

   126 135 132 116

Calculate the mean monthly figure, using an assumed mean.

5. The heights of a class of boys, measured to the nearest cm, are shown below. Using an assumed mean, calculate the mean height.

Height (cm)	155	156	157	158	159	160	161	162	163	164	165	166	167
Frequency	1	1	5	2	3	3	5	2	4	0	1	1	1

6. The length of the right foot of each girl in a third year class was measured to the nearest half centimetre. Calculate the mean length of foot (using an assumed mean of 22.0 cm).

Length (cm)	20.0	20.5	21.0	21.5	22.0	22.5	23.0	23.5	24.0	24.5	25.0	25.5	26.0
Frequency	0	2	4	2	4	4	3	3	2	1	0	0	1

7. Repeat the last question, by measuring the right foot of all the girls *or* the boys in your own class. (Why should you not take the boys and girls together for this?)

8–11. Calculate the means for the questions in Exercise C using the assumed mean method.

The median from a frequency distribution

It is very easy to find the median of a set of raw scores by setting the scores out in numerical order and counting along till the middle is found.

The same method is used to find the median of a frequency distribution when the data is ungrouped.

Example
Find the median of the following scores.

Score	51	52	53	54	55	56
Frequency	3	4	7	6	3	1

Solution
There is a total frequency of 24, so the median must lie between the 12th and 13th scores. We examine the frequencies to see where the 12th and 13th scores lie. These scores are both 53, so the median must be 53.

When the data we are concerned with is grouped into class intervals, it is only possible from the distribution, to say that the median lies in a certain group. To find the median more exactly we draw the cumulative frequency curve for the distribution and estimate the median from this.

Example
Find the median of the marks gained by 50 pupils in an English examination.

Mark	21–30	31–40	41–50	51–60	61–70	71–80
Frequency	2	6	11	20	7	4

Solution
Construct a cumulative frequency table and draw the curve.

Mark	Cum. Freq.
21–30	2
21–40	8
21–50	19
21–60	39
21–70	46
21–80	50

Fig. 5.1 Cumulative Frequency Curve of Examination Marks

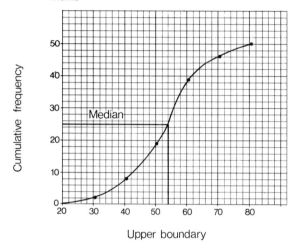

Fig. 5.2 (a) Geography Examination Results

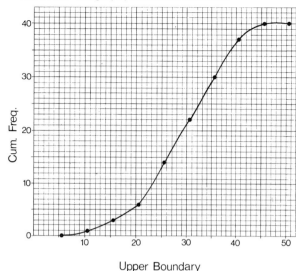

2. The curve in Figure 5.2b, shows the marks of the same class for a history exam.

a) Estimate the median mark from the graph.

b) If the pass mark was 24 what percentage of the class failed?

c) What percentage of the class passed?

There is a total frequency of 50, so the median lies between the 25th and 26th score i.e. it is the $25\frac{1}{2}$th score.

We draw a line across the graph at the $25\frac{1}{2}$th score until it cuts the curve and then drop another line to the horizontal axis. Thus from the graph the median is the mark of 54.

Note: If the total frequency is N, then the

median is $\frac{N}{2} + \frac{1}{2}$

Where N is large the $\frac{1}{2}$ becomes insignificant and the median is taken as $\frac{N}{2}$.

EXERCISE E

1. The frequency curve (Figure 5.2a) shows the marks of 40 pupils in a geography exam.

a) From the curve, estimate the median mark.

b) If 25% of the pupils failed, what was the pass mark?

Fig. 5.2 (b) History Examination Results

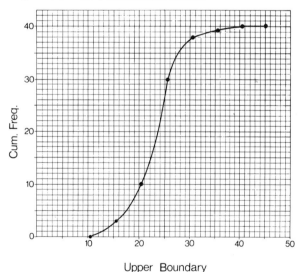

3–6. Find the cumulative frequency curves you drew for the questions in Exercise H, Chapter 4 and estimate the medians for each distribution.

7. Shown in the table on the right are the number of deaths registered in Wales for for each week of a recent year.

Construct a cumulative frequency table, draw the cumulative frequency curve and from it estimate the median number of deaths registered.

Number of Deaths	Frequency
450–499	3.
500–549	16
550–599	16
600–649	11
650–699	3
700–749	2
750–799	1

Comparing the three 'averages'

The mean

Advantages
1. It can be calculated exactly.
2. It makes use of all the data.
3. It can be used in further statistical calculations.

Disadvantages
1. It can be very misleading if there is an abnormally high or low value, e.g. the ages of 10 pupils in a school are 5, 5, 5, 6, 6, 6, 7, 7, 7, 17 years. The mean age is 7.1 years. In this case 9 of the 10 pupils are below the mean age. The abnormally high value of 17 has unduly affected the mean.

The median

Advantages
1. It is simple to understand.
2. It is unaffected by abnormally high or low values.
3. It is characteristic of the normal group and sometimes represents an actual member of the group e.g. the pupil of median height in a class can actually be picked out and examined.

Disadvantages
1. It cannot be used in further statistical calculations.
2. Its value can only be estimated (from a cumulative frequency curve) in grouped distributions.
3. In small groups or in groups which have a rather odd pattern of distribution, it may not be characteristic of the group e.g. in a test there were 6 scores of 5 marks, 2 of 7 marks, 1 of 8 marks and 1 of 9 marks. From these marks the median would be 5 which would not be a good average to use for the group. (In this case, the mean would be a better average to use.)

The mode

Advantages
1. It is simple to understand.
2. It is unaffected by abnormally high or low values.
3. It is the average useful to manufacturers of shoes, clothes, hats etc.

Disadvantages
1. It cannot be determined exactly in a distribution where the data is grouped.
2. It cannot be used in arithmetical calculations.

54

The geometric mean

Another average which is sometimes used in statistics is the geometric mean.

The geometric mean of n numbers is the nth root of their product.

e.g. The geometric mean of 3, 6 and 9
$$= \sqrt[3]{(3 \times 6 \times 9)}$$
$$= \sqrt[3]{162} = 5.5$$
The geometric mean of 3, 6, 9 and 10
$$= \sqrt[4]{(3 \times 6 \times 9 \times 10)}$$
$$= \sqrt[3]{1620} = 6.3$$

The calculation of the cube root, 4th root, 5th root etc. of numbers may be done using logarithms or by using a scientific calculator.

EXERCISE F

Calculate the geometric mean of the following sets of numbers (correct to one decimal place).

1. 2, 18
2. 2, 3
3. 2, 3, 5
4. 2, 7, 8
5. 3, 4, 5, 6
6. 4, 6, 8, 10
7. 2, 3, 4, 5, 6
8. 3, 5, 6, 7, 8
9. 2, 3, 5, 7, 9, 10
10. 1, 2, 3, 5, 7, 9

EXERCISE G Miscellaneous

1. Find the mean, median and mode of the following sets of measures.
   a) 3, 3, 5, 5, 5, 7, 7.
   b) 61, 67, 76, 90, 90, 90.
   c) 24, 24, 25, 31, 40, 50, 63, 79.

2. Find the geometric mean of each of the following sets of measures.
   a) 2, 6, 18 b) 5, 10, 15, 20 c) 3, 5, 7, 9, 11

3. A pupil scored the following set of marks in a series of tests.
$$6, 4, 4, 3, 7, 9, 8, 4, 5, 5, 6$$
   a) Determine the mean, median and mode of these marks.
   b) Which of these measures would the pupil prefer to tell his parents was his average?

4. 600 items were examined and the number of faults per item was found to be as follows.

No. of faults	0	1	2	3	4	5	6
Frequency	34	132	185	95	75	64	15

   a) State the modal number of faults.
   b) Calculate (correct to one decimal place) the mean number of faults per item.
   c) Find the percentage of items which were perfect.
   d) Find the percentage of items with more than two faults.

5. The length of time, in minutes, spent in the waiting room of a hospital by 50 patients was as follows.

```
35 73 46 32 26 49 83 65 29 74
61 36 52 42 32 57 67 47 51 53
27 61 49 36 29 28 55 57 65 61
55 42 54 36 51 32 60 57 47 58
47 61 60 54 61 28 60 42 43 61
```

   a) Group the data in classes 26–35, 36–45 etc.
   b) Calculate the mean length of time spent in the waiting room.
   c) Construct the cumulative frequency distribution, draw the cumulative frequency curve and thus obtain the median time spent in the waiting room.
   d) What percentage of the patients spent more than one hour in the waiting room?

6. The weights of 40 lambs were taken and the results are tabulated below.

Weight (kg)	11–14	15–18	19–22	23–26	27–30	31–34
No. of lambs	0	8	16	11	4	1

   a) Find the mean weight.
   b) Draw the cumulative frequency curve and find the median weight.

c) What is the lower limit of the weights included in the modal class?

d) What percentage of the lambs were less than 25 kg?

7. The first 30 words of a book were taken and the number of letters per word recorded as follows.

No. of letters per word	1	2	3	4	5	6	7	8	9	10	11
No. of words	1	5	8	7	2	3	2	0	0	1	1

a) What is the modal number of letters?

b) Calculate the mean number of letters per word.

c) Find the median number of letters per word.

8. For the following distributions which of the three statistics — mean, median or mode — has the smallest value?

9. a) Use the cumulative frequency curve drawn in Figure 4.11 to find the median mark for the distribution.

b) Use the cumulative frequency curve drawn in Figure 4.12 to find the median mark for the distribution.

c) Use the cumulative frequency curve drawn in Figure 4.13 to find the median length of a trout.

10. Use the data given in Chapter 4, Exercise J, question 3 to find the mean length of time between scheduled take-offs.

11. Use the information given in Chapter 4, Exercise J, question 4 to find the mean weight of the a) girls, b) boys in the sample.

12. Use the data given in Chapter 4, Exercise J, question 6 to find

a) the mean lifetime of the electric bulbs,

b) the median lifetime of the electric bulbs.

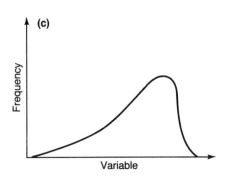

Fig. 5.3

6 APPLICATIONS OF THE MEAN

Moving averages

A **time series** graph is a line graph drawn from a series of readings taken over regular time intervals, perhaps every month, every quarter-year or every year.

Very often graphs of this type demonstrate a **cyclic** or seasonal variation.

The graph shown below for the sales of ice-cream recorded every quarter-year illustrates this quite clearly. Over each year, the sales for the various quarters exhibit the same relative increases and decreases.

The irregularity of this type of graph disguises the general "trend" of data. It is very useful to add a **trend line** to the graph

	Time	Sales (£ thousands)
1978	1st quarter	10
	2nd quarter	20
	3rd quarter	25
	4th quarter	12
1979	1st quarter	10
	2nd quarter	28
	3rd quarter	38
	4th quarter	16
1980	1st quarter	14
	2nd quarter	38
	3rd quarter	48
	4th quarter	26
1981	1st quarter	24
	2nd quarter	46

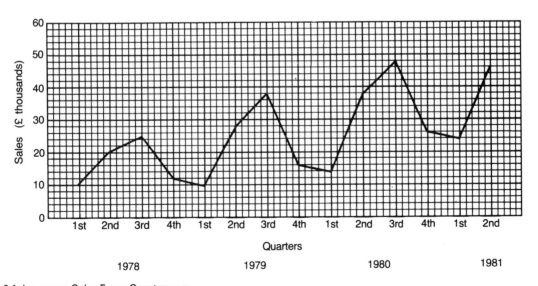

Fig. 6.1 Ice-cream Sales Every Quarter-year

which as the name suggests shows the overall trend of the data. To draw a trend line, the **moving averages** are calculated. When we are dealing with quarterly figures a four-point moving average is used. The first average is the average of the figures for the four quarters of 1978. The second average is the average of the 2nd, 3rd and 4th quarters of 1978 and the 1st quarter of 1979. The third average is the average of the 3rd and 4th quarters of 1978 and the 1st and 2nd quarters of 1979. Further averages are obtained in succession by omitting the oldest measure and taking into account one new measure each time. We continue in this manner until we have reached the last four measures.

The necessary calculations for the moving averages can be tabulated as shown.

Note particularly that the moving totals and the moving averages are listed opposite the middle point of the time interval concerned. In the table, the first moving total of 67 and the average of 17 are set against the midway point between the 2nd and 3rd quarter.

Similarly, when these averages are added to the graph, the points are plotted midway between the quarters. The average points are joined to give the trend line.

Time	Sales £ (thousands)	Moving annual total	Moving average
1978			
1st quarter	10		
2nd quarter	20	67	17
3rd quarter	25	67	17
4th quarter	12	75	19
1979			
1st quarter	10	88	22
2nd quarter	28	92	23
3rd quarter	38	96	24
4th quarter	16	106	27
1980			
1st quarter	14	116	29
2nd quarter	38	126	32
3rd quarter	48	136	34
4th quarter	26	144	36
1981			
1st quarter	24		
2nd quarter	46		

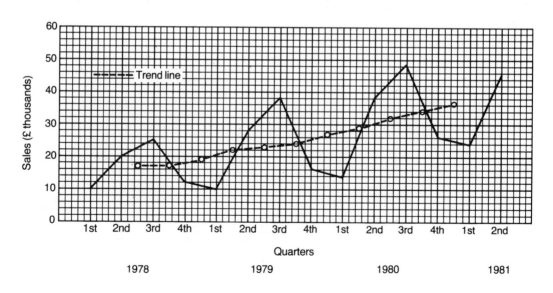

Fig. 6.2 Trend line of Ice-cream Sales

58

In our example graph the trend line shows that there has been a steady increase in the sales of ice-cream.

In this example since four measurements were averaged it is said to be a four-point moving average, but if the data had been given in terms of monthly sales figures then it would have been appropriate to average successive sets of 12 measures and this would be called a 12-point moving average. When seasonal variations are to be smoothed out then it is essential that sets of measures for complete seasons be taken for the calculation of the moving averages.

Moving averages can be used to smooth out a graph when the irregularities are the result of other than seasonal variations. The only change is that since there is no complete season available as the obvious size of group for the calculation of the averages an arbitrary choice must be made in the matter.

Example
The number of passes in higher mathematics in a Scottish Secondary School for each of the last 15 years are as shown at the top of the next column.

Calculate the three-point moving averages for these measures and illustrate the raw

Year	No. of Passes	Year	No. of Passes
1957	35	1965	44
1958	33	1966	48
1959	47	1967	68
1960	43	1968	56
1961	39	1969	65
1962	28	1970	72
1963	33	1971	74
1964	32		

measures and their moving averages on the same grid.

Solution

(Figure 6.3 and following table)

It can be noted that, in this case, the graph is only partially smoothed out by taking a three-point moving average. A more regular curve could be obtained by using, say, a five-point moving average.

In general moving averages are used to smooth out a graph or distribution of measures thereby eliminating the variations and bringing out the general trend of the measures. If the variations are seasonal then

Fig. 6.3

——————— Raw measures
.... Trend line

Year	No. of Passes	Three year Moving Totals	Three-point Moving Averages
1957	35		
1958	33	115	38
1959	47	123	41
1960	43	129	43
1961	39	110	37
1962	28	100	33
1963	33	93	31
1964	32	109	36
1965	44	124	41
1966	48	160	53
1967	68	172	57
1968	56	189	63
1969	65	193	64
1970	72	211	70
1971	74		

the group chosen for averaging should cover a complete season otherwise the size of group is an arbitrary choice.

EXERCISE A

1. Calculate the three-point moving averages of the following measures:

    20   15   5   10   12   8   15   25   10   15

2. Calculate the five-point moving averages of the measures quoted in question 1.
3. In a school in Scotland the number of passes in Higher French for each of the last 12 years was as follows:

Year	No. of Passes	Year	No. of Passes
1960	21	1966	26
1961	23	1967	32
1962	25	1968	23
1963	23	1969	31
1964	31	1970	39
1965	24	1971	35

Calculate the four-point moving averages for these measures and illustrate the actual number of passes and the moving averages on the same grid.

4. Consumer expenditure on cars and motor cycles, in millions of pounds, for each quarter of the years 1967 to 1970 were as listed below.

	1967	1968	1969	1970
1st quarter	150	153	155	183
2nd quarter	180	161	182	206
3rd quarter	216	189	200	219
4th quarter	228	209	187	237

Calculate the four-point moving averages for these measures and illustrate the raw data and the moving averages on the same grid.

5. The number of live births in England and Wales for the last two quarters of 1968 and for each quarter of the years 1969 and 1970 are listed below. Calculate the four-point moving averages for these measures and illustrate the raw measures and the moving averages on the same grid.

Live Births Registered (in thousands)			
	1968	1969	1970
1st quarter		213.1	196.3
2nd quarter		206.1	204.1
3rd quarter	207.7	199.0	196.5
4th quarter	190.3	180.0	187.6

6. The number of passenger movements by air and sea both into and out from the United Kingdom for each quarter of the years 1968 to 1970 are tabulated below.

1968	Inward		Outward	
	Sea	Air	Sea	Air
1st quarter	362	1058	352	1040
2nd quarter	1112	2119	1195	2146
3rd quarter	2643	3194	2577	3211
4th quarter	502	1451	496	1440
1969				
1st quarter	420	1237	419	1241
2nd quarter	1323	2527	1415	2534
3rd quarter	2932	3753	2912	3785
4th quarter	566	1728	563	1697
1970				
1st quarter	552	1494	589	1544
2nd quarter	1406	2929	1443	2891
3rd quarter	3144	4345	3103	4381
4th quarter	1533	2046	626	2029

Calculate four-point moving averages for:

    a) sea movements inward
    b) sea movements outward
    c) air movements inward
    d) air movements outward

7. The daily takings in a small gift shop for the five weeks prior to Christmas are as follows. Calculate the six-point moving averages for the takings and plot the graph of these moving averages. (Takings are given correct to the nearest pound.)

	Week 1	Week 2	Week 3	Week 4	Week 5
	£	£	£	£	£
Mon.	36	40	46	45	45
Tue.	53	59	68	75	71
Wed.	46	39	35	50	39
Thur.	62	75	92	104	87
Fri.	85	94	138	152	121
Sat.	120	132	175	215	165

8. The total numbers of unemployed persons in the development areas, expressed as a percentage of the total population for the years 1958 to 1967 are listed below. Calculate the three-point moving averages for these measures.

Year	% unemployed	Year	% unemployed
1958	3.3	1963	4.2
1959	3.7	1964	3.1
1960	2.9	1965	2.4
1961	2.4	1966	2.2
1962	3.4	1967	3.5

9. The gross profits of companies as a percentage of their total income for the years 1959 to 1970 are as follows:

Year	% Profit	Year	% Profit
1959	15.1	1965	14.5
1960	15.5	1966	13.0
1961	14.3	1967	13.2
1962	13.7	1968	13.4
1963	14.7	1969	12.5
1964	15.1	1970	11.4

Calculate the four-point moving averages.

Weighted mean

In some situations it is unsatisfactory to calculate the ordinary mean of a set of measures. The situation may be such that greater emphasis must be given to particular measures. This is commonly done by using a weighted mean. The technique is best illustrated by example.

Example 1
To estimate the likely performance of a class in a national examination the pupils sat two separate tests which were both marked out of 100. The teacher considered that the second test was twice as valuable as the first. If a pupil scored 40% in the first test and 55% in the second then on this basis his weighted mean is obtained by averaging his first result with twice the second result.
i.e. Weighted Mean =

$$\bar{x}_w = \frac{40 + 2 \times 55}{1 + 2} = \frac{150}{3} = 50.$$

Example 2
If in the above the decision had been to weight the results in the ratio 2:3 then the Weighted Mean =

$$\bar{x}_w = \frac{2 \times 40 + 3 \times 55}{2 + 3} = \frac{245}{5} = 49.$$

Example 3
A manufacturer makes three models A, B and C on which he reckons to make a profit of £50, £40, £30 respectively. His results show that for every 2 models of type A which he sells he is able to sell 3 models of type B and 5 of type C. Find his mean profit per article sold.
Weighted Mean =

$$\bar{x}_w = \frac{2 \times £50 + 3 \times £40 + 5 \times £30}{2 + 3 + 5}$$

$$= \frac{£370}{10}$$

$$= £37$$

In general if the measures $x_1, x_2, x_3, \ldots x_n$ are given weights $w_1, w_2, w_3, \ldots w_n$ then their

Weighted Mean =

$$\bar{x}_w = \frac{w_1 x_1 + w_2 x_2 + w_3 x_3 + \ldots + w_n x_n}{w_1 + w_2 + w_3 + \ldots + w_n}$$

EXERCISE B

1. Find the weighted mean of two measures 5 and 10 if they are given weights of 2 and 3 respectively.
2. Find the weighted mean of the measures 10, 14, 15, 18, 20 if they are given weights of 1, 2, 3, 4, 5 respectively.
3. In the three sections of an examination a candidate scored 40%, 45% and 60%. Find the weighted mean if the three sections are given weights 1, 2 and 2 respectively.
4. An alloy is made by mixing three parts, by volume of material A with five parts, by volume of material B. If the density of material A is 5 gm per cm^3 and that of material B is 9 gm per cm^3 find the density of the alloy.
5. The two isotopes of chlorine have atomic masses of 35 and 37. If the two isotopes are normally found in a mixture in the ratio of three of the first to one of the second find the average atomic mass of the mixture.
6. The two isotopes of copper have atomic masses of 63 and 65. If these isotopes are normally found in a mixture in the proportion of 7 of the first to 3 of the second find the average atomic mass of the mixture.

Index numbers

Two measures can be compared in a variety of ways:

1) the two measures can be stated
2) the difference between the measures can be given
3) the one measure can be given as a fraction of the other
4) the percentage change can be given
5) the one measure can be given as a percentage of the other.

The last of these methods of comparison is the concept underlying an index number. The only fundamental difference is that in an index number the percentage sign is omitted. Index numbers are at present normally used to express changes in prices and the figure upon which the percentage is calculated is called the base.

Example 1

An article which cost £5 in 1968 now costs £7. Find the index number corresponding to the present price if the 1968 price is taken as base.

$$\text{Index number} = \frac{7}{5} \times 100 = 140.$$

Example 2

In 1968 the index number corresponding to the cost of an article on the 1965 cost as base was 120. The present index number taken on the 1968 cost as base is 105. Find the index number for the present cost taken on the 1965 cost as base.

Index number = 105% of 120

$$= \frac{105}{100} \times 120 = 126$$

Example 3

If the index number for an article in 1968 was 120 and at present is 150, both index numbers being taken on the 1965 cost as base, find the index number for the present cost taken on the 1968 cost as base.

$$\text{Index number} = \frac{150}{120} \times 100 = 125.$$

EXERCISE C

1. Calculate the index numbers corresponding to the 2nd cost of each of the following commodities taking the first cost as base in each case:

Commodity	1st cost £	2nd cost £
A	2.00	2.10
B	5.50	6.60
C	2.50	2.00
D	34.50	53.20*
E	245	452*

2. In each of the following calculate the present cost if the given index number is the present one and is calculated on the first cost as base:

Commodity	1st cost £	Index number
A	10	105
B	7.80	120
C	0.84	125
D	345	97†
E	345	103†

3. In each of the following examples the 1970 and 1971 index numbers are calculated on the 1968 cost as base. Express the 1971 cost as an index number calculated on the 1970 cost as base.

Commodity	1970 Index No.	1971 Index No.
A	110	132
B	120	144
C	120	108
D	90	72
E	123	134*

4. The index number for an item in 1968 was 105, calculated on the 1966 cost as base, and in 1970 the index number for the same item calculated on the 1968 cost as base was 140. Calculate the index number for the 1970 cost on the 1966 figure as base.

*Correct to the nearest integer

†Correct to the nearest £

5. The index number of an article in 1970 was 90, calculated on the 1969 cost as base and the index number of the same article in 1971 was 110, calculated on the 1970 cost as base. Find the index number in 1971 for that article calculated on the 1969 cost as base.

Index of retail prices

This measure is frequently quoted in the press and changes in this measure are frequently used as grounds for claims for wage increases. Fundamentally this index is a combination of the ideas behind weighted mean and index numbers. A large representative sample of the population has been required to keep a very careful note of all forms of spending incurred by the family over a period of time and from the analysis of all of this information details of the spending of the average family have been obtained. Decisions have then been taken as to the classification of all of the items of spending and weights have been allocated to these classes of spending in proportion to the results obtained from the survey. The total of the weights allocated is normally 1000. The pattern of family spending changes as national habits change and thus the allocation of these weights is kept continually under review.

Some particular date is chosen as a base and each class of goods is allocated the index number 100 for that date. Changes in the cost of items within each class can be kept under review and appropriate index numbers can be obtained, at any time, for the various classes of expenditure. The weighted mean of these various index numbers can be calculated and this measure is called the index of retail prices. The full list of details of the index of retail prices is too long to be quoted here but a summary of the facts given as at June 1970 is as follows: (January 1962 = 100)

	Weight	Index
Food	298	142
Tobacco and Alcoholic drink	130	140
Housing, Fuel and Light	180	153
Household Goods	60	125
Clothing	86	123
Transport	126	131
Services	55	152
Miscellaneous	65	142
Total	1000	

Example 1

The cost of producing an article is weighted as follows: labour 55, materials 25, plant and maintenance and depreciation 20. On the 1969 costs as base the index numbers for these sections of the cost are 160, 135 and 110 respectively. Find the index number corresponding to the present cost of production on the basis of the 1969 costs.

Solution

	Weight	Index Number
Labour	55	160
Materials	25	135
Plant	20	110
Total	100	

Present Index Number

$$= \frac{55 \times 160 + 25 \times 135 + 20 \times 110}{55 + 25 + 20}$$

$$= \frac{14375}{100}$$

$$= 144$$

Example 2

If the 1968 index of retail prices had the following weights and indices as based on the 1962 prices find the 1968 index of retail prices.

64

Solution

	Weight	Index Number
Food	304	124
Tobacco and Alcohol	129	126
Housing and Durable Goods	331	127
Transport	120	119
Misc. Goods and Services	116	128
Total	1000	

Index of Retail Prices

$$= \frac{304 \times 124 + 129 \times 126 + 331 \times 127 + 120 \times 119 + 116 \times 128}{304 + 129 + 331 + 120 + 116}$$

$$= \frac{125115}{1000}$$

$$= 125$$

EXERCISE D

1. A shopkeeper estimates his expenditure under the undernoted headings and allocates the accompanying weights and index numbers to the items taking his 1965 costs as base. Find the weighted index number corresponding to these details.

	Weight	Index Number
Wages	140	170
Cost of Goods	765	155
Rent and Rates	35	210
Maintenance of Premises	60	110

2. As part of the evidence for a claim for increased pocket money the boys of a boarding school produced the following details regarding the weighting of their spending and the changes in costs over the last year.

	Weight	Index Number
Sweets	25	125
Comics	10	130
Pictures	35	110
Toys etc.	20	120
Savings and presents	10	120

Find the weighted index number corresponding to these measures and state the percentage increase in pocket money that they would be justified in claiming on this evidence.

3. A housewife studied her outlay on food etc. and allocated weights to the various classes of expenditure as below. From a comparison of prices of the item involved she estimates the attached index numbers, using the 1969 prices as base. Calculate the weighted index number for the range of expenditure.

	Weight	Index Number
Groceries	42	105
Meat and fish	20	115
Fruit and vegetables	8	150
Milk and eggs	12	106
Bread and cakes	10	110
Cleaning materials	8	105

4. Calculate the weighted index number corresponding to the following items and their attached weights and index numbers.

	Weight	Index Number
Food	316	121
Tobacco and Drink	125	110
Housing	160	115
Clothing	83	109
Durable Goods	80	105
Transport	120	112
Services	60	115
Misc. Goods	56	110

5. Calculate the index number for all the groups listed below by using the given data.

	Weight	Index Number
Food	395	118
Tobacco and Drink	115	102
Rent and Rates	75	115
Fuel and Light	62	114
Clothing	83	98
Durable Goods	55	96
Transport	105	112
Misc. Goods and Services	110	100

7

MEASURES OF DISPERSION

Range

In Chapter 5, we studied measures of central tendency or averages. These averages are very important since they can give us a picture of the group which they represent. However, taken by themselves, they give us a very limited view of the whole picture. As well as the average, we need to know how the rest of the data is grouped around the average—whether it is closely gathered round or scattered more widely. In other words we have to find some way of measuring how the scores scatter. We need a measure of the spread, scattering or dispersion of the scores.

Examine the data shown below of the weekly wages of ten employees in different factories.

In each factory the mean wage is £30. But the way the wages are scattered about the mean is very different in each case. One simple way of measuring the scatter is to consider the **range** of values. In some situations this is helpful but in the above example, the range is the same in factory

A, B and C. As a measure of dispersion the range is very easily found but is not particularly good. Thus more satisfactory measures are required.

Interquartile range

In many experimental situations we tend to distrust extreme measures and as a result these measures are often discarded. It is this type of thinking that leads to improved versions of the range called the **interquartile range** and the **semi-interquartile range**.

Just as the median divides a set of scores into two equal parts, the **quartiles** divide a set of scores into four equal parts. There are three quartiles, the lower, the middle and the upper quartiles.

The lower quartile is called Q_1
The middle quartile (usually called the median) is Q_2
The upper quartile is Q_3
Then the Interquartile Range is $Q_3 - Q_1$
and the Semi-Interquartile Range is $\dfrac{Q_3 - Q_1}{2}$

Weekly Wage	Employees in Factory A	Employees in Factory B	Employees in Factory C	Employees in Factory D
£50	2	5	1	0
£40	2	0	2	4
£30	2	0	4	2
£20	2	0	2	4
£10	2	5	1	0

Example 1
Find the upper and lower quartiles and the interquartile range of this set of scores.
$$8, 9, 7, 10, 5, 4, 6$$

Solution
First write out the scores in numerical order.

4	5	6	7	8	9	10
	Q_1		Q_2		Q_3	

Fix the position of the median first, then since there are three scores below the median and three scores above the median the lower quartile is the second score and the upper quartile is the sixth score.

$$\text{Thus } Q_1 \text{ (lower quartile)} = 5$$
$$Q_3 \text{ (upper quartile)} = 9$$
and the Interquartile Range $= 9 - 5 = 4$.

Example 2
Find the range and the semi-interquartile range of the following set of scores.
$$12, 12, 3, 8, 9, 9, 5, 6, 3, 4, 5, 11$$

Solution
First arrange the scores in numerical order.

3	3	4	5	5	6	8	9	9	11	12	12
		$Q_1 = 4.5$			$Q_2 = 7$			$Q_3 = 10$			

Then locate the median and since there are six scores below and above the median the lower quartile is the average of the third and fourth scores and the upper quartile is the average of the ninth and tenth scores.

$$\text{Thus the median} = \frac{6+8}{2} = 7$$

$$\text{The lower quartile} = \frac{4+5}{2} = 4.5$$

$$\text{The upper quartile} = \frac{9+11}{2} = 10$$

$$\text{The semi-interquartile range} = \frac{10-4.5}{2} = 2.75$$

$$\text{The range} = 12-3 = 9$$

The quartiles from a frequency distribution

We find the quartiles from a frequency distribution in the same way as we did for the median in chapter 5 i.e. we estimate the quartiles from the cumulative frequency curve of the distribution. When dealing with frequency distributions with a small number of measures we shall again consider the number of measures above and below the median as an aid to the location of the quartiles.

Example
The cumulative frequency table for the marks gained by 50 pupils in an English examination is shown opposite. Since there are 50 measures there are 25 measures above and below the median and thus the lower quartile corresponds to the 13th score and the upper quartile corresponds to the 38th score.

Mark	Cum. Freq.
21–30	2
21–40	8
21–50	19
21–60	39
21–70	46
21–80	50

Figure 7.1 (a) shows the cumulative frequency curve for the distribution with the median and quartiles shown.

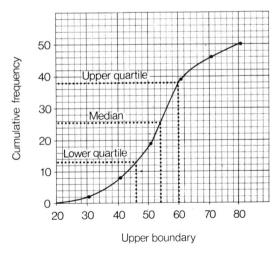

Fig. 7.1 (a) Cumulative Frequency Curve for the Distribution with the Median and Quartiles shown

67

From the graph,

the median	$= 54$
the lower quartile	$= 46$
the upper quartile	$= 60$

and thus the semi-interquartile range

$$= \frac{60 - 46}{2}$$

For large values of the total cumulative frequency, N, the fine details covered in the above discussion become rather insignificant and we can say that

the lower quartile corresponds to $\frac{1}{4}N$
the median corresponds to $\frac{1}{2}N$
the upper quartile corresponds to $\frac{3}{4}N$

N–large

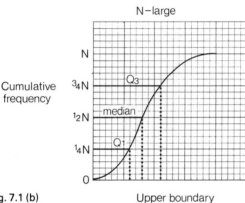

Fig. 7.1 (b)

Interpercentile range

Just as the quartiles divide a distribution into four equal parts, the percentiles divide a distribution into one hundred equal parts.

Obviously,

the lower quartile is the 25th percentile,
the median is the 50th percentile

and, the upper quartile is the 75th percentile.

Again, as with the quartiles, we estimate the percentiles from a cumulative frequency curve.

Using percentiles we have another measure of the dispersion of a distribution called the interpercentile range. This is the range between the 10th percentile and the 90th percentile. Again, as with the interquartile range it does not depend on the extreme values, but as with the range, the inter-quartile and semi-interquartile ranges it does not lend itself to further mathematical treatment.

An easy way to obtain any percentile is to draw another axis to the right of the graph of the cumulative frequency curve and mark the scale from 0 to 100, making 100 level with the total frequency.

If we want to find, say, the 10th and 90th percentile we draw lines across to the curve as shown below and so obtain the values of the variable.

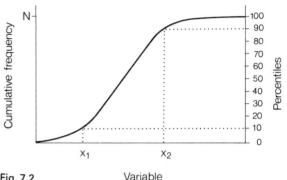

Fig. 7.2   Variable

EXERCISE A

1. Find the range and the semi-interquartile range of the following sets of scores.

    a) 20, 18, 17, 23, 19, 21.
    b) 11, 12, 13, 15, 7, 6, 5, 10, 14, 8, 9.
    c) 108, 108, 110, 101, 110, 111, 104, 104, 105, 107.
    d) 4, 5, 5, 1, 1, 2, 8, 8, 9, 9, 8, 7, 2.

2.

Weight (kg) *(Upper Boundary)*	*Cumulative Frequency*
66	5
67	8
68	12
69	20
70	35
71	47
72	59
73	68
74	72
75	75

This cumulative frequency table shows the weight in kg of 75 senior boys. Draw the

cumulative frequency curve and from it estimate:
  a) the median
  b) the interquartile range
  c) the 10–90 percentile range
  d) how many boys weigh 69.5 kg or less
  e) how many boys weigh more than 71.5 kg.

3. 140 pupils sat two examinations in Arithmetic. The frequency table shows the distribution of their marks.

Marks	1st Exam. Freq.	2nd Exam. Freq.
0–10	0	4
11–20	3	6
21–30	10	9
31–40	34	9
41–50	42	12
51–60	32	20
61–70	12	42
71–80	5	22
81–90	2	10
91–100	0	6

Construct cumulative frequency tables and draw the cumulative frequency curves for each examination on the same graph.
  a) Estimate the medians
  b) Find the semi-interquartile ranges
  c) If the pass mark for the first examination was 45, how many pupils passed?
  d) What percentage of the pupils failed this examination?
  e) If 78 pupils failed in the second examination, what was the pass mark?
  f) In a few words compare the two distributions and comment on the standard of the test papers.

4. These are the heights (to the nearest cm) of some boys and girls of the same age.

Boys' heights

159	162	166	158	162	164	158	170
165	165	165	159	162	147	147	160
162	163	164	161	164	162	163	165
163	161	164	161	165	167	162	159
162	167	165	158	166	164	158	160
163	160	161	164	162	162		

Girls' heights

162	157	161	164	162	159	160	164
163	156	165	162	163	161	160	162
164	162	161	162	159	160	162	162
161	160	164	163	164	162	167	161
161	162	159	164	162	165	161	161
158	159	157	165	161	163		

Form a frequency table for the two distributions using groups starting at 147–148, 149–150 etc.
  Construct a cumulative frequency table and draw the two curves on the same graph.
  a) Find the median for each distribution.
  b) Find the interquartile ranges.
  c) How many girls are 159 cm or less in height?
  d) How many boys are 159 cm or less in height?
  e) How many girls are taller than 163 cm?
  f) How many boys are taller than 163 cm?
  g) Comment on the two distributions.

5. 60 pupils sat an examination in English and one in Arithmetic and the results are summarised below:

Mark	11–20	21–30	31–40	41–50	51–60	61–70	71–80	81–90
Frequency (English)	1	4	10	15	17	9	3	1
Frequency (Arithmetic)	0	1	2	20	32	4	1	0

By drawing cumulative frequency curves estimate the median, quartiles and the semi-interquartile range for each of the distributions.

Mean deviation

We require a measure of dispersion or variability which takes account of all the measures of the distribution. One such measure is the **mean deviation.** This, as it implies, is the mean of the deviations or differences of the scores from either the mean, median or mode. It is usually most useful to calculate the mean deviation of the scores from the mean since the mean itself depends on all the measures in the distribution.

The mean deviation is calculated as shown below in the example. Since the negative signs must be ignored, any further algebraic development of the concept is made impossible, though the mean deviation would be quite satisfactory for many purposes.

Example
Calculate the mean deviation, from the mean, of the following scores.

10   12   14   16   18   20   22

Solution

$$\text{The mean, } \bar{X} = \frac{\sum X}{X} = \frac{112}{7} = 16$$

Score X	Deviations from mean $\bar{X} = 16$	Deviations taken as positive
10	−6	6
12	−4	4
14	−2	2
16	0	0
18	+2	2
20	+4	4
22	+6	6
		$\sum d = 24$

If the deviations were summed as they stand in the second column the answer would be 0, so all the deviations are taken as positive.

The sum of the deviations, $\sum d = 24$

$$\text{and the mean of the deviations} = \frac{\sum d}{N}$$

$$= \frac{24}{7}$$

$$= 3.43$$

EXERCISE B

Find the mean deviation, from the mean, of these sets of scores.
1. 1, 3, 4, 6, 7, 9, 12
2. 60, 63, 65, 66, 69, 70, 76
3. 1, 5, 11, 13, 15, 17, 20, 22

Standard deviation

If we add two extra steps to the calculation which was performed to find the mean deviation we obtain a measure of dispersion called the **standard deviation.** This is the most satisfactory measure of dispersion, since it makes use of all the scores in a distribution and is also quite acceptable mathematically.

In calculating the standard deviation, the mean of the distribution is found, the deviations of the scores from the mean are tabulated, these deviations are squared (thus dealing with the negative signs in an acceptable algebraic manner), and the mean of the squares of the deviations is calculated. The square root of this mean is the standard deviation.

Thus,

$$\text{the Standard Deviation} = \sqrt{\frac{\sum (X - \bar{X})^2}{N}},$$

where X stands for the various measures, $\bar{X}$ = the mean and N = the number of measures in the distribution.

Note: If we stop at finding the mean of the squares of the deviations, we have the **variance**

$$\text{i.e. the Variance} = \frac{\sum (X - \bar{X})^2}{N}$$

To calculate the standard deviation

To calculate the standard deviation of the scores 10, 11, 12, 13, 14, 15, 16 the following method is used.

1. List the scores in order.

2. Calculate the mean $\overline{X} = \dfrac{\sum X}{N}$

$$= \dfrac{91}{7}$$

$$= 13$$

3. List the deviations of each score from the mean.

Score X	Deviation of score from mean $X-\overline{X}$	Squares of deviations $(X-\overline{X})^2$
10	$10-13 = -3$	9
11	$11-13 = -2$	4
12	$12-13 = -1$	1
13	$13-13 = 0$	0
14	$14-13 = +1$	1
15	$15-13 = +2$	4
16	$16-13 = +3$	9

$$\sum(X-\overline{X})^2 = 28$$

4. Square the deviations (this eliminates the negative signs).
5. Sum the squares of the deviations $\sum(X-\overline{X})^2 = 28$.
6. Divide this by the number of scores, N = 7.
7. Take the square root of the answer, which is the standard deviation

$$s \text{ (standard deviation)} = \sqrt{\dfrac{\sum(X-\overline{X})^2}{N}}$$

$$= \sqrt{\dfrac{28}{7}}$$

$$= \sqrt{4}$$

$$= 2$$

Example 1
Calculate the mean and standard deviation of the following scores.

22 24 26 28 30 32 34 36

Solution

Score (X)	$X-\overline{X}$	$(X-\overline{X})^2$
22	$22-29 = -7$	49
24	$24-29 = -5$	25
26	$26-29 = -3$	9
28	$28-29 = -1$	1
30	$30-29 = +1$	1
32	$32-29 = +3$	9
34	$34-29 = +5$	25
36	$36-29 = +7$	49

$$\sum(X-\overline{X})^2 = 168$$

$$\overline{X} = \dfrac{\sum X}{N}$$

$$= \dfrac{232}{8}$$

$$= 29$$

$$s = \sqrt{\dfrac{\sum(X-\overline{X})^2}{N}}$$

$$= \sqrt{\dfrac{168}{8}}$$

$$= \sqrt{21} \text{ (Use logarithms or square root tables)}$$

$$= 4.58$$

Mean = 29

Standard Deviation = 4.58

Example 2
Calculate the mean and standard deviation of this frequency distribution of test marks.

Mark	0	1	2	3	4	5	6	7	8	9	10
Frequency	1	2	3	5	6	7	4	3	2	1	0

Solution
With a frequency distribution we use exactly the same process as in example 1 but to the table we must add a frequency column, an fX column and an $f(X-\overline{X})^2$ column.
The formula used now becomes

$$s = \sqrt{\dfrac{\sum f(X-\overline{X})^2}{\sum f}}$$

71

Mark (X)	f	fX	$(X-\bar{X})$	$(X-\bar{X})^2$	$f(X-\bar{X})^2$
0	1	0	$0-4.5 = -4.5$	20.25	20.25
1	2	2	$1-4.5 = -3.5$	12.25	24.50
2	3	6	$2-4.5 = -2.5$	6.25	18.75
3	5	15	$3-4.5 = -1.5$	2.25	11.25
4	6	24	$4-4.5 = -0.5$	0.25	1.50
5	7	35	$5-4.5 = +0.5$	0.25	1.75
6	4	24	$6-4.5 = +1.5$	2.25	9.00
7	3	21	$7-4.5 = +2.5$	6.25	18.75
8	2	16	$8-4.5 = +3.5$	12.25	24.50
9	1	9	$9-4.5 = +4.5$	20.25	20.25
10	0	0	$10-4.5 = +5.5$	30.25	0
	$\sum f = 34$	$\sum fX = 152$			$\sum f(X-X)^2 = 150.50$

$$\bar{X} = \frac{\sum fX}{\sum f}$$

$$= \frac{152}{34}$$

$$= 4.5 \text{ (correct to 1st decimal place)}$$

$$s = \sqrt{\frac{\sum f(X-\bar{X})^2}{\sum f}}$$

$$s = \sqrt{\frac{\sum f(X-\bar{X})^2}{\sum f}}$$

$$= \sqrt{\frac{150.5}{34}}$$

$$= \sqrt{4.43} \text{ (correct to 2nd decimal place)}$$

$$= 2.10$$

Mean = 4.5

Standard deviation = 2.10

Note: When working with a grouped distribution use the mid-point of the interval to stand for the group and proceed as above.

An alternative method for calculating the standard deviation

The above method for calculating the

standard deviation works fairly well (example 1) when the mean is a whole number and there is no frequency. However, when the mean is not a whole number and there is a frequency involved and possibly grouped data also, the arithmetical calculation is rather excessive.

From the original formula, the following formula may be derived.

$$\text{Standard Deviation} = \sqrt{\frac{\sum X^2}{N} - \left(\frac{\sum X}{N}\right)^2}$$

and with a frequency distribution,

$$\text{Standard Deviation} = \sqrt{\frac{\sum fX^2}{\sum f} - \left(\frac{\sum fX}{\sum f}\right)^2}$$

This derived formula simplifies the calculation considerably.

Look again at the worked example.

Score X	X^2
10	100
11	121
12	144
13	169
14	196
15	225
16	256
$\sum X = 91$	$\sum X^2 = 1211$

$$\text{Mean} = \frac{\sum X}{N} = \frac{91}{7} = 13$$

$$s = \sqrt{\frac{\sum X^2}{N} - \left(\frac{\sum X}{N}\right)^2}$$

$$= \sqrt{\frac{1211}{7} - 13^2}$$

$$= \sqrt{173 - 169}$$

$$= \sqrt{4} = 2$$

Look now at example 2.

Mark X	f	fX	fX²
0	1	0	0
1	2	2	2
2	3	6	12
3	5	15	45
4	6	24	96
5	7	35	175
6	4	24	144
7	3	21	147
8	2	16	128
9	1	9	81
10	0	0	0
	$\sum f = 34$	$\sum fX = 152$	$\sum fX^2 = 830$

$$\bar{X} = \frac{\sum fX}{\sum f} = \frac{152}{34} = 4.47$$

$$s = \sqrt{\frac{\sum fX^2}{\sum f} - \left(\frac{\sum fX}{\sum f}\right)^2}$$

$$= \sqrt{\frac{830}{34} - (4.47)^2}$$

$$= \sqrt{24.41 - 19.98} = \sqrt{4.43} = 2.10$$

It seems clear that this method is simpler than the basic one for use with a frequency distributions, especially when the mean is not exact.

Note: To get the last column in the table "fX²", the fX column is multiplied by X again i.e. fX × X = fX².

EXERCISE C

1. Calculate the mean and standard deviation of each of the following sets of scores.
   a) 3, 4, 5, 6, 7, 9, 10. 12.
   b) 24, 29, 27, 20, 20, 23, 21.
   c) 66, 66, 60, 64, 69, 67, 63.
   d) 35, 37, 37, 40, 30, 34, 33, 38.
   e) 30, 33, 24, 28, 20, 17, 25, 39, 34, 42.

2. In ten successive years in Glasgow the highest temperature (in the shade) for each year was recorded as:

   78, 82, 82, 80, 79, 76, 76, 78, 72, 74. (°F)

   Calculate the mean and standard deviation of these temperatures.

3. In the same ten years, the lowest temperature (in the shade) recorded each year was:

   12, 24, 15, 18, 12, 15, 18, 11, 19, 11. (°F)

   Calculate the mean and standard deviation of the temperatures.

4. The number of days in which rain fell (rain days) each month in 1964 were recorded as follows.

Month	North Scotland	East Scotland	West Scotland
Jan.	23	18	22
Feb.	19	15	18
Mar.	19	16	17
Apr.	19	16	17
May	17	15	16
June	17	14	16
July	19	17	18
Aug.	19	17	19
Sept.	21	16	19
Oct.	22	19	21
Nov.	22	18	20
Dec.	22	18	21

For each region calculate the mean and standard deviation of the number of rain days.

Compare briefly the three distributions explaining any differences or similarities.

5. Below are the heights of 40 fourth year boys, measured to the nearest 2 cm. Calculate the mean height and the standard deviation of the heights.

Height (cm)	150	152	154	156	158	160	162	164	166	168	170	172	174	176	178	180
Frequency	1	1	2	3	4	4	5	6	4	4	1	2	2	0	0	1

6. These are the heights of 30 fourth year girls measured at the same time as the boys.

Height (cm)	146	148	150	152	154	156	158	160	162	164	166	168	170
Frequency	1	2	2	3	3	7	2	5	1	2	1	0	1

Calculate the mean and standard deviation of the girls' heights and compare them to your answers for question 5. Give reasons for any differences you find.

7. Measure the heights of a number of boys or girls in your school and calculate the mean and standard deviation.

8. The frequency table below shows the distribution of marks in an examination.
   Calculate the mean and standard deviation of the marks.

Mark	1–5	6–10	11–15	16–20	21–25	26–30	31–35	36–40	41–45	46–50
Frequency	0	2	4	11	20	10	5	3	2	1

9. The weights of 40 fourth year boys measured to the nearest kilogramme, are shown below. Calculate the mean and standard deviation of the weights.

Weight (kg)	35–39	40–44	45–49	50–54	55–59	60–64	65–69	70–74	75–79	80–84
Frequency	1	6	9	9	6	4	3	0	1	1

10. Take the weights of 40 fourth year boys in your school and calculate the mean and standard deviation. Compare your results with those of question 9.

11. This table shows the death-rates per 1000 population of a number of countries for 1964.

Form a frequency table grouping the data, starting at 6.0—6.9, 7.0—7.9 etc. Calculate the mean and standard deviation of the death-rates for these countries.

Country	Death-rate per 1000
Eastern Germany	13.5
Austria	12.3
Belgium	12.1
Scotland	11.7
Eire	11.3
England and Wales	11.0
Western Germany	10.7
France	10.5
Northern Ireland	10.2
Portugal	10.0
Denmark	10.0
Norway	10.0
Sweden	10.0
Hungary	9.9
Czechoslovakia	9.6
Italy	9.6
U.S.A.	9.4
Yugoslavia	9.4
Finland	9.3
Switzerland	9.2
Australia	9.0
New Zealand*	8.8
Spain	8.7
Netherlands	7.7
Canada	7.6
Poland	7.6
Japan	6.9
Israel†	6.0

* excluding Maoris
† Jewish population only.

12. The times (measured to the nearest second) taken by rats to go through a maze in a laboratory test, are shown below. Construct a frequency distribution and calculate the mean and standard deviation of the times.

8 26 12 12 18 15 12 14 16 20 19
14 14 11 17 22 9 21 13 17 24 18
20 17 19 13 18 13 16 20 10 18 24
20 15 26 23 15 19 21 23 18 22 25
25 19 23 16 28 21

(Group in threes starting 8–10.)

Calculation of the standard deviation from an assumed mean

The calculation of the standard deviation, as discussed in the last section, may be further simplified by the use of an assumed mean. This is specially useful when the values of the variable are quite high.

The formula used now is—

$$\text{Standard Deviation} = \sqrt{\frac{\Sigma d^2}{N} - \left(\frac{\Sigma d}{N}\right)^2}$$

Where d is the deviation of each measure from the assumed mean.

When this formula is extended to cover a frequency distribution it becomes—

$$\text{Standard Deviation} = \sqrt{\frac{\Sigma fd^2}{\Sigma f} - \left(\frac{\Sigma fd}{\Sigma f}\right)^2}$$

Finally, if we have a grouped frequency distribution and we use the group width as the unit of measure of deviation from the assumed mean, then the last formula would require to be multiplied by the group width. Thus if C is the group width,

$$\text{Standard Deviation} = C\sqrt{\frac{\Sigma fd^2}{\Sigma f} - \left(\frac{\Sigma fd}{\Sigma f}\right)^2}$$

Example 1
Calculate the mean and standard deviation of
the distribution of heights below:

Height (cm)	162	163	164	165	166	167	168	169	170	171	172
Frequency	1	3	9	13	18	20	12	11	7	4	2

Solution

X	f	X_0	d	fd	fd^2
162	1		−5	−5	25
163	3		−4	−12	48
164	9		−3	−27	81
165	13		−2	−26	52
166	18		−1	−18	18
				−88	
167	20	167	0	0	0
168	12		1	+12	12
169	11		2	+22	44
170	7		3	+21	63
171	4		4	+16	64
172	2		5	+10	50
				+81	$\sum fd^2 = 457$
				$\sum fd = -7$	

$$\text{Mean} = 167 + \frac{(-7)}{100} = 166.9 \text{ cm (to 4 sig. fig.)}$$

$$\text{Standard Deviation} = \sqrt{\frac{\sum fd^2}{\sum f} - \left(\frac{\sum fd}{\sum f}\right)^2}$$

$$= \sqrt{\frac{457}{100} - \left(\frac{-7}{100}\right)^2}$$

$$= \sqrt{4.57 - 0.0049}$$

$$= \sqrt{4.565} = 2.13 \text{ cm}$$

Example 2
Calculate the mean and standard deviation
of the following distribution of class marks.

Mark	10–19	20–29	30–39	40–49	50–59	60–69	70–79	80–89
Frequency	1	5	7	9	18	14	4	2

Solution

X	f	X_0	d	fd	fd^2
10–19	1		−4	−4	16
20–29	5		−3	−15	45
30–39	7		−2	−14	28
40–49	9		−1	−9	9
				−42	
50–59	18	54.5	0	0	0
60–69	14		+1	+14	14
70–79	4		+2	+8	16
80–89	2		+3	+6	18
	$\sum f = 60$			+28	$\sum fd^2 = 146$
				$\sum fd = -14$	

(Note that the unit of measure of d is the group width)

$$\text{Mean} = 54.5 + \frac{(-14) \times 10}{60}$$

$$= 54.5 - 2.3 = 52.2$$

$$\text{Standard Deviation} = C\sqrt{\frac{\sum fd^2}{\sum f} - \left(\frac{\sum fd}{\sum f}\right)^2}$$

$$= 10\sqrt{\frac{146}{60} - \left(\frac{14}{60}\right)^2}$$

$$= 10\sqrt{2.433 - 0.054}$$

$$= 10\sqrt{2.378}$$

$$= 10 \times 1.54$$

$$= 15.4$$

1. The heights of 26 pupils, measured to the nearest cm, are tabulated below. Find the mean and standard deviation of the distribution.

Height (cm)	148	149	150	151	152	153	154	155	156
Frequency	1	1	2	3	6	7	3	2	1

2. Find the mean and standard deviation of the following distribution of scores:

Score	51	52	53	54	55	56
Frequency	3	4	7	6	3	1

3. The rateable values of eighty houses were found, to the nearest pound, and the results are shown in the following table. Find the mean and standard deviation of the distribution.

Rateable value in pounds	50–59	60–69	70–79	80–89	90–99	100–109	110–119
Frequency	2	10	20	24	12	11	1

4. Calculate the mean and standard deviation of the following distribution of class marks.

Mark	1–5	6–10	11–15	16–20	21–25	26–30	31–35	36–40	41–45
Frequency	1	3	5	7	13	9	7	3	2

5. Calculate the mean and standard deviation of each of the distributions of marks tabulated in Chapter 7, Exercise A, Question 3.

6. Calculate the mean and standard deviation of each of the distributions of heights tabulated in Chapter 7, Exercise A, Question 4.

7. Calculate the mean and standard deviation of each of the distributions of marks tabulated in Chapter 7, Exercise A, Question 5.

Calculation of the mean and standard deviation of a distribution derived from another distribution

If we have a distribution with a known mean and standard deviation which is altered by an arithmetical operation, it is possible to deduce the mean and standard deviation of the new distribution without re-calculation.

Suppose we have a distribution ranging from 9 to 23 with a mean of 16 and a standard deviation of 2. If 10 is added to each score, the whole distribution moves up the number line but the "spread" of the scores remains the same.

So, the new mean is $16 + 10 = 26$ and the new standard deviation is still 2.

If, however, each score in the original distribution is multiplied by 2, this time the "spread" of the scores is also multiplied by 2.

The new mean is $16 \times 2 = 32$ and the new standard deviation is $2 \times 2 = 4$.

EXERCISE E

1. A set of marks has a mean of 15 and a standard deviation of 2.
    Write down the new mean and standard deviation if each mark—
    a) is multiplied by 3,
    b) is doubled,
    c) has 10 added to it,
    d) is increased by 5,
    e) is halved,
    f) has 3 subtracted from it.

2. A set of scores has a mean of 100 and a standard deviation of 9.
    Write down the new mean and standard deviation if every score—
    a) is quartered,
    b) is multiplied by 3,
    c) has 10 subtracted from it,
    d) is increased by 20,
    e) is doubled and then increased by 5,
    f) is halved and then decreased by 5.

3. A set of marks has a mean of 23 and a standard deviation of 2.5.
    Write down the new mean and standard deviation if each mark—
    a) is multiplied by 5,
    b) has 1 added to it,
    c) has 3 subtracted from it,
    d) is halved,
    e) is doubled and has 2 added on,
    f) is multiplied by 3 and has 2 subtracted from it,
    g) has 2 added on and is then doubled,
    h) has 1 subtracted from it and is then halved.

4. A set of measures has a mean of 52 and a standard deviation of 7.8.
    Write down the new mean and standard deviation if every measure—
    a) is doubled and has 1 added on,
    b) is halved and has 1 subtracted,
    c) is increased by 5 and then doubled,
    d) has 2 subtracted from it and is then halved,
    e) has 10 added on and is then multiplied by 3.

5. 2, 5, 7, 7, 8, 9, 11.

For this set of figures, calculate—
    a) The mean.
    b) The standard deviation.
    c) If each figure is doubled and then increased by 5, find the new mean and standard deviation.

6. 50, 52, 53, 55, 57, 59, 62, 64, 65, 66.

For this set of figures, calculate—
    a) The mean.
    b) The standard deviation.
    c) If every figure is increased by 10 and then halved, find the new mean and standard deviation.
    d) If every figure is quartered and then has 2 subtracted from it, find the new mean and standard deviation.

Standard scores

The comparison of measures from different distributions is usually unsatisfactory and the addition of such results frequently produces unexpected results. These difficulties are particularly obvious in the comparison of school examination results and when we add these marks to produce an average performance. The surprising results are usually caused by an unusual mean or dispersion in one or more of the distributions. These difficulties can be overcome if we use what are commonly called **standard** or **standardised scores.** When using standard scores we wish to eliminate the effect of the mean and the dispersion and the first of these is done by subtracting the mean of the distribution from each score. The deviate so obtained is then divided by the standard deviation to remove the effect of the dispersion. The result of this division is called the standard score. Standard scores from different distributions can then be compared or combined without being affected by either bias.

Example 1
A boy scored 50 marks in an English examination where the class mean was 45 and the standard deviation 5. He scored 60 marks in an Arithmetic examination where the class mean was 55 and the standard deviation 10. Compare his standing in the two subjects.

Solution

Standard Score in English $= \dfrac{50-45}{5} = \dfrac{5}{5} = +1$

Standard Score in Arithmetic $= \dfrac{60-55}{10} = \dfrac{5}{10} = +\frac{1}{2}$

Thus the boy did better in English than Arithmetic, compared to the rest of the class.

Example 2
Tom, Dick and Harry sat examinations in Maths, English and French. Their scores, the class means and standard deviations for each subject are shown in the following table.

	Maths	English	French	Total
Tom	81	50	46	177
Dick	42	57	55	154
Harry	33	56	65	154
Class Mean	48	52	50	
Standard Deviation	15	4	10	

Using the given raw scores Tom would be placed first. By using standard scores check the validity of this placing.

Solution

Standard Scores

	Maths	English	French	Total
Tom	$\dfrac{81-48}{15} = +2.2$	$\dfrac{50-52}{4} = -0.5$	$\dfrac{46-50}{10} = -0.4$	$+1.3$
Dick	$\dfrac{42-48}{15} = -0.4$	$\dfrac{57-52}{4} = +1.25$	$\dfrac{55-50}{10} = +0.5$	$+1.35$
Harry	$\dfrac{33-48}{15} = -1.0$	$\dfrac{56-52}{4} = +1.0$	$\dfrac{65-50}{10} = +1.5$	$+1.5$

The correct order of merit is Harry, Dick, Tom.

EXERCISE F

1. Jane scored 66 marks in her Geography examination in November where the mean was 60 and the standard deviation 6. In May, when the mean was 40 and the standard deviation 10, she had a mark of 55 for her Geography. In which examination did she show her ability better?

2. The following table shows the marks of a boy in three successive mathematics examinations.

Exam.	Mark	Mean	Standard Deviation
1st exam.	50	53	3
2nd exam.	51	58	7
3rd exam.	53	48	5

a) In which examination was his relative standing best?
b) Was his performance better in the second examination than in the first?

3 a). In an Arithmetic test given to a class, the mean and standard deviation were 64 and 8 respectively. The table below shows the marks of some of the pupils. Complete the table.

Pupil	Mark	Deviate from mean	Standard Score
Alan	68	$68 - 64 = 4$	$+\frac{1}{2}$
Betty	72	$72 - 64 = 8$	$+1$
Colin	56		
David	60		
Evelyn	80		
Fay	76		

b). In another Arithmetic test given to the same class a few weeks later, the mean was 49 and the standard deviation 6. Complete the following table.

c). i) Which of the six pupils improved most?

ii) Which pupil came down most in the second test?

Pupil	Mark	Deviate from mean	Standard Score
Alan	58		
Betty	52		
Colin	43		
David	49		
Evelyn	64		
Fay	55		

iii) Which pupil was most consistent in the two tests?

4. In the first term John scored 94% in Maths and Mary scored 80% while in the second term John scored 78% and Mary scored 90%. If the mean scores were first term: 50%, and second term: 60%, and the standard deviations were first term: 20%, and second term: 15%, find which pupil had the better average performance.

5. Mary, Jean and Helen sat examinations in English, French and Latin. The results and the class means and standard deviations are tabulated below. Find the correct order of merit.

	English	French	Latin
Mary	59	73	79
Jean	58	81	75
Helen	64	74	74
Class Mean	52	60	65
Standard deviation	5	10	8

EXERCISE G   Miscellaneous

1. a) Calculate the mean and standard deviation of the following set of measures.
    10, 11, 12, 13, 14, 15, 16
b) Without repeating the calculation, state the mean and standard deviation of each of the following sets of measures.
   1) 12, 13, 14, 15, 16, 17, 18
   2) 20, 22, 24, 26, 28, 30, 32
   3) 22, 24, 26, 28, 30, 32, 34
c) The measures in a) are scaled to give a mean of 50 and a standard deviation of 12. Find the new values of the measures.

2. For the set of measures
$$2, 3, 5, 7, 8, 10, 14$$
calculate—
  a) the median,
  b) the interquartile range,
  c) the semi-interquartile range,
  d) the range,
  e) the mean,
  f) the mean deviate from the mean,
  g) the standard deviation.
  h) If each measure is doubled and then increased by 5, state the new mean and standard deviation.

3. For the set of measures
$$9, 10, 12, 15, 18, 20, 22, 23, 24$$
calculate—
  a) the range,
  b) the median,
  c) the interquartile range,
  d) the mean,
  e) the mean deviate from the mean,
  f) the standard deviation.
  g) If each measure has 2 subtracted from it and is then halved, state the new mean and standard deviation.

4. For the set of measures
$$17, 14, 8, 5, 9, 13, 22, 20$$
calculate—
  a) the range,
  b) the median,
  c) the interquartile range,
  d) the mean,
  e) the standard deviation.
  f) If each measure is multiplied by 5 and then has 10 subtracted from it, write down the new mean and standard deviation.

5. For the set of scores
$$3, 4, 4, 11, 9, 3, 8, 7, 5$$
calculate—
  a) the mode,
  b) the range,
  c) the median,
  d) the interquartile range,
  e) the semi-interquartile range,
  f) the mean,
  g) the mean deviate from the mean,

h) the standard deviation.
i) If each score has 5 added to it and is then divided by 3, write down the new mean and standard deviation.

6. The mean and standard deviation of a set of scores are 50 and 12 respectively. The measures are altered by scaling so that a raw score of 44 becomes a scaled score of 55.
  a) If the mean of the scaled scores is 60 find their standard deviation.
  b) If the standard deviation of the scaled scores is 20 find their mean.

7. Find the mean deviate from the mean of the following sets of measures:
  a) 9, 11, 12, 13, 14, 15, 17.
  b) 18, 22, 24, 26, 28, 30, 34.

c)

Measure	0	1	2	3	4	5
Frequency	11	14	10	5	3	1

d)

Measure	1–5	6–10	11–15	16–20	21–25
Frequency	1	3	5	7	13

Measure	26–30	31–35	36–40	41–45
Frequency	9	7	3	2

8. For each of the distributions given in question 7 calculate the variance.

9. For the distribution of measures given in question 7(d) draw a cumulative frequency curve and use it to find—
  a) the median,
  b) the semi-interquartile range
  c) the interpercentile range.

10. The yield of a group of 80 fruit trees was measured correct to the nearest kilogramme. The results of the survey are given below.

Yield (kg)	10–14	15–19	20–24	25–29	30–34	35–39
No. of Trees	3	18	27	22	6	4

  a) Estimate the median yield.
  b) Find the interquartile range of the distribution.

c) Find the mean yield.

d) Find the standard deviation of the distribution.

11. The table given below shows the marks gained by 30 pupils in a test.

Mark	1	2	3	4	5	6	7	8	9	10
Frequency	0	1	0	3	7	6	5	3	4	1

Calculate the mean and standard deviation of these results. What is the modal mark?

12. The age distribution of workers in a factory was as follows.

Age in years	Frequency
16–20	2
21–25	10
26–30	12
31–35	23
36–40	17
41–45	13
46–50	9
51–55	6
56–60	4
61–65	4

a) Calculate the mean and standard deviation of this distribution.

b) Draw the cumulative frequency curve for this distribution and use it to find the median, interquartile and interpercentile ranges.

13. Find the mean and standard deviation of the following set of measures.

Length (cm)	1	2	3	4	5	6	7
Frequency	1	4	5	7	1	1	1

Use the above result to obtain the mean and standard deviation of the following distribution.

Weight (kg)	61	62	63	64	65	66	67
Frequency	5	20	25	35	5	5	5

14. Calculate the mean and standard deviation of the distribution.

Measure	50–59	60–69	70–79	80–89	90–99	100–109
Frequency	3	10	18	24	20	5

15. To calculate the mean and standard deviation of a set of marks, 60 was subtracted from each mark and then the mean and standard deviation of the new scores were found to be 4 and 12 respectively.

a) What was the mean of the original set of marks?

b) What was the standard deviation of the original set of marks?

16. A group of candidates for a post were given three tests. After two tests, candidates X, Y and Z appear to be the best of the group. Their results in these tests and the means and standard deviations are tabulated below.

Test	Mean	Standard Deviation	Scores		
			X	Y	Z
1	70	6	79	85	80
2	55	10	70	62	68

At this stage, which candidate has the best total standard score? In the third test where the mean is 40 and the standard deviation is 12, Y has a score of 58. If X beat Y on total standard score by 0.3, what was X's raw score in the third test?

8 REGRESSION AND CORRELATION

Regression and correlation

In this study, to date, there has only been one form of measure under consideration at any one time. In this chapter the interest is in situations where two forms of measure are being considered at the same time e.g. the height and weight of a group of children, attainment in arithmetic and the intelligence quotient of a group of 10 year old children. The incentive to compare measures in pairs as suggested above usually arises from a belief that the measures are in some way linked. The aim of this chapter is to find a method of illustrating the measures so that their interdependence, if it exists, can be seen, to measure the amount of their interdependence and use the given data to estimate other results concerning the given situation.

Scatter diagrams

The standard method of illustrating this type of situation is by using a scatter diagram and this consists of laying out each of the measures along one of the axes of a grid, then taking each item, in turn, the two measures for that item act exactly like an ordered pair and thus like the coordinates of a point on the grid. Each item considered is thereby linked to one point on the grid and that point can be plotted in the normal way and is usually marked on the grid as an 'X'. The points corresponding to all of the items being measured are similarly located on the grid and plotted and thus a scatter of 'Xs' is built up on the grid.

Example
The reading speeds in words per minute and the intelligence quotients of a group of 12 children were measured and are tabulated below.

The scatter diagram (Figure 8.1) is built up as follows. The axes are set out to cover a suitable range of measures of Reading Speed and Intelligence Quotient. Each child is then considered in turn and the two measures e.g. for the child A, Reading Speed 120 and Intelligence Quotient 80, locate one point on the grid and this point is marked with a dot. Similar dots are entered in turn for the other children and the complete scatter diagram will consist of 12 dots.

Child	A	B	C	D	E	F	G	H	I	J	K	L
Reading Speed. in words per minute	120	140	100	170	130	190	220	140	180	240	200	270
Intelligence Quotient	80	90	90	100	100	105	110	110	115	120	125	130

Fig. 8.1 Reading Speed (in Words per Minute)

From an examination of the pattern of crosses obtained it can be accepted that the above results help to support the belief that there is some linkage between Reading Speed and Intelligence Quotient but since there is some scattering of the points plotted indications are that the two measures are not very closely linked. It must also be accepted that no conclusive results could ever be obtained from such a small sample of the population.

The distribution of points plotted on a scatter diagram can give an indication of the relation which exists between the two characteristics being measured. Three basic forms of scatter diagram are shown below. In Figure 8.2a the points plotted are scattered at random over the grid and this indicates that the measures are not related. Figure 8.2b shows the other extreme position as here the points are scattered along a straight line and

this is the typical diagram for measures which are directly proportional to each other. In a scatter diagram like Figure 8.2c the closer the points are to being on a straight line the greater the degree of their possible linkage.

For measures to be related it is not necessary that the scatter diagram approximates to a straight line through the origin or even to any straight line. If the points plotted on the scatter diagram fit some straight line or curve this can be taken as evidence for some relation between the measures and it then becomes necessary to put the relation into algebraic form.

Regression

The results of scientific experiments can often be illustrated by the use of scatter diagrams and if these results indicate that the measures under consideration are related it is then

Fig. 8.2

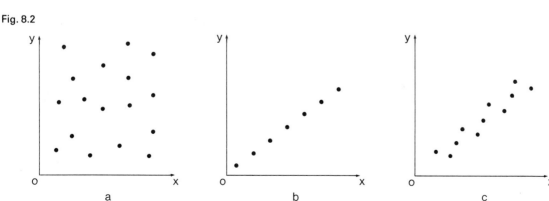

85

desirable that the results are interpreted in a mathematical form. If the points plotted are almost on a straight line then the line which most nearly fits all of the points must be drawn. This line is called the 'Best Fitting Line' or the 'Regression Line'. If the measures are such that the points plotted are almost exactly on a straight line then there are few complications about the drawing of the line which most nearly fits all of the points but if the points are more scattered it then becomes possible to draw two regression lines. If the variables are, for convenience, called x and y then the one line is called the regression line of x on y and in it the values of y are assumed to be accurate and the points deviate from the straight line because of errors in the measurement of x. These deviations in the values of x will hereafter be called offsets. The method of drawing such a regression line will be discussed in the next section but having drawn this regression line it could be used to estimate the value of x which would correspond to a given value of y. The other regression line, called the regression line of y on x is drawn in a similar fashion except that the roles of x and y are interchanged.

Drawing of regression lines (best fitting lines)

There are two main methods of drawing regression or best fitting lines.

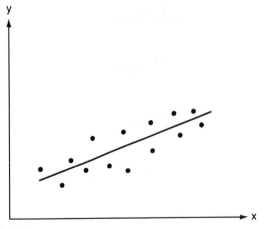

Fig. 8.3

86

1) *By inspection*
   This means that, without calculation, a line is drawn which appears to look the "best fitting" line.

2) *By the method of semi-averages*
   In this method, the scatter diagram is "halved" by drawing in the median line for the x-values. Then the mean value of x and y for each half is calculated and these two points plotted. When a straight line is drawn through these points, the line of best fit is obtained.

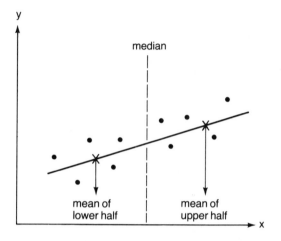

Fig. 8.4

Example
The marks of 8 pupils from tests in Maths and Science are shown below.

Pupil	1	2	3	4	5	6	7	8
Maths	5	6	10	11	14	15	18	23
Science	5	14	9	16	19	14	18	25

Draw a scatter diagram using the Maths marks for the x-axis and the Science marks for the y-axis, and draw in the line of best fit by the method of semi-averages. Using this line find the average Science mark of pupils whose Maths mark is 20 (assuming that the group of pupils tested is representative of all pupils)

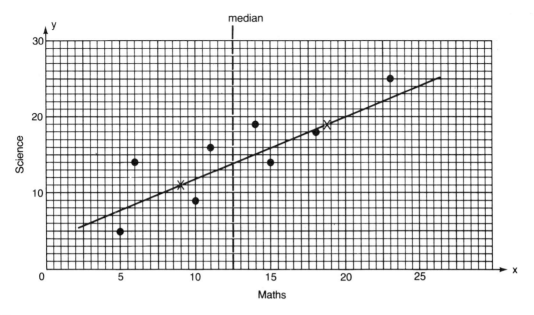

Fig. 8.5

Solution
Once the diagram (Figure 8.5) is drawn and the points plotted, the line showing the median of the x-values is marked in.
Lower half—

$$\Sigma x = 32, \text{ the mean of x} = \frac{\Sigma x}{N} = \frac{32}{4} = 8$$

$$\Sigma y = 44, \text{ the mean of y} = \frac{\Sigma y}{N} = \frac{44}{4} = 11$$

This gives us a mean point (8,11) to plot.

Upper half—

$$\Sigma x = 70, \text{ the mean of x} = \frac{\Sigma x}{N} = \frac{70}{4} = 17\tfrac{1}{2}$$

$$\Sigma y = 76, \text{ the mean of y} = \frac{\Sigma y}{N} = \frac{76}{4} = 19$$

This gives us a mean point $(17\tfrac{1}{2},19)$ to plot.
The straight line joining these two points (the means of the lower and upper halves of the graph) gives the best fitting line.
This line can now be used to find the average y-value for a given x-value. It should be noted that this average value refers to the underlying population and not to the sample.

So, for a Maths mark of 20, the average Science mark would be 21.

EXERCISE A

1. The following figures show the results of tests in English and French given to 10 pupils. Draw a scatter diagram to illustrate the marks, taking the English marks for the x-axis and the French marks for the y-axis. Use a scale of 2 cm for 10 marks on both axes.
   Using the method of semi-averages draw the best fitting line.
   Find the average mark in French for pupils getting a mark in English of
   a) 50 *b)* 11 *c)* 70.

Pupil	A	B	C	D	E	F	G	H	I	J
English	10	14	20	23	28	35	38	43	47	52
French	8	15	21	12	24	16	29	22	27	36

2. The marks below for Algebra and Geometry are the result of tests given to eight pupils. Draw a scatter diagram to show them, taking the Algebra marks for the x-axis (scale, 5 marks for 2 cm) and the Geometry

87

marks for the y-axis (scale, 10 marks for 2 cm). Using the method of semi-averages, draw the line of best fit.

Use this line to find the average mark in Geometry for pupils getting a mark in Algebra of

　　a) 16　　　　*b)* 25　　　　*c)* 50.

Pupil	1	2	3	4	5	6	7	8
Algebra	9	14	16	19	22	27	30	33
Geometry	10	14	20	16	22	29	28	35

3. A group of pupils were given two tests in Arithmetic with a time gap of two months between the tests. The results for eight of these pupils are shown below. Draw a scatter diagram to illustrate these results, taking Test 1 as the x-axis (2 cm for 10 marks) and Test 2 as the y-axis (2 cm for 10 marks). Draw the best fitting line using the method of semi-averages.

From this line, find the average mark of pupils in Test 2 who scored a mark in Test 1 of

　　a) 37　　　　*b)* 50　　　　*c)* 10.

Test 1	24	30	32	40	43	47	55	63
Test 2	31	25	28	34	39	33	37	47

4. The points awarded by two judges to six competitors in an ice-skating competition are shown in the table below. Plot the points to obtain a scatter diagram, taking the marks of the 1st Judge for the x-axis (scale for both axes — start at 4.5 and take 1 cm for every 0.1 of a point). Using the method of semi-averages, draw the line of best fit on the diagram.

What score (to the nearest 0.1) would be given by the 2nd Judge when the 1st Judge awards

　　a) 5.2　　　　　　　*b)* 5.5?

Skater	1	2	3	4	5	6
1st Judge	5.9	5.0	5.5	4.8	5.3	5.7
2nd Judge	5.8	5.1	5.2	5.1	5.5	5.6

5. A group of pupils were given tests in French and German. The results are shown in the table. Draw a scatter diagram to illustrate them (French for x-axis, German for y-axis). Use the method of semi-averages to draw in the best fitting line.

Use this line to find the average German mark for a French mark of

　　a) 11　　　　*b)* 30　　　　*c)* 17.

Pupil	1	2	3	4	5	6	7	8
French	10	31	37	11	26	18	15	22
German	9	34	35	15	26	20	14	21

6. Two successive tests in English were given to a group of pupils. The marks of six of the pupils are shown below. Illustrate the marks in a scatter diagram (Test 1 x-axis, Test 2 y-axis) and draw in the line of best fit using the method of semi-averages.

Estimate the mark gained in the second test by a pupil who scored in the first test a mark of

　　a) 25　　*b)* 40　　*c)* 45　　*d)* 10.

Pupil	1	2	3	4	5	6
Test 1	25	42	50	30	61	32
Test 2	20	27	35	15	40	22

7. Some children were tested to find their scores in an I.Q. test and their marks in a Maths test. The results are shown in the table. Use these results to draw a scatter diagram taking the I.Q. scores for the x-axis. Using the method of semi-averages, draw the best fitting line.

Use this line to estimate the Maths mark for a child with an I.Q. score of

　　a) 113　　　　*b)* 140　　　　*c)* 131.

I.Q. Score	96	98	104	110	117	123
Maths mark	27	20	50	38	57	69
I.Q. Score	126	128	130	132	135	141
Maths mark	54	63	78	89	70	93

8. Some engineering apprentices were given two tests in Maths at an interval of two months.

Test 1	35	80	93	85	61	47	41	74
Test 2	40	75	79	80	54	54	42	66

Draw a scatter diagram, and draw in the line of best fit, using the method of semi-averages.

What mark in the second test would be expected if the marks in the first test were
   a) 50   *b*) 70   *c*) 90?

9. The same boys were given two tests in technical drawing at an interval of one month.

Test 1	10	60	60	20	34	25	53	47
Test 2	11	45	35	25	14	25	32	36

Draw a scatter diagram to show these results and, using the method of semi-averages, draw in the best fitting line. Estimate the marks gained in the second test for first test marks of
   a) 20   *b*) 42   *c*) 59.

Equation of the regression line

Once the regression line has been drawn it is

Fig. 8.6

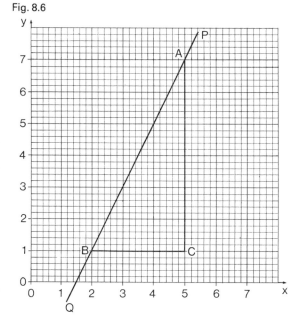

usually necessary that the equation of the line be expressed in mathematical terms. Consider the line PQ shown on the following diagram.
On the line select two points A and B which are well separated, one near each end of the line. From these points A and B draw lines AC and BC parallel to the axes in order to complete the right angled triangle ABC. The standard form of the equation of a straight line is $y = mx + c$ where m is the gradient of the line and c is a constant. If the coordinates of the points A and B are represented by the symbols $A(x_A, y_A)$ and $B(x_B, y_B)$ then the gradient of $AB = m$ is given by the fraction $\frac{CA}{BC} = \frac{y_A - y_B}{x_A - x_B}$. In the context of regression this fraction also gets the name of the **coefficient of regression.**
Once the value of m has been established the constant c can be obtained by the substitution of the coordinates of either A or B into the equation $y = mx + c$. A good choice of the points A and B can help to reduce the weight of calculation. The points should be chosen so that the coordinates can be read without the use of too many decimals or fractions and since the calculation involves division by BC it is desirable that the length BC be chosen so as to make for an easy division.

Example
Considered the regression line drawn in Figure 8.5 and find the equation of the regression line. Also, use the equation to estimate the value of y which would correspond to 1) $x = 12$, 2) $x = 22$.
Since the line was drawn by the method of semi-averages we can use the points through which the line was drawn for the calculation i.e. we can take $A(17\frac{1}{2}, 19)$ and $B(8, 11)$ and let the equation of the line be of the form $y = mx + c$, then

$$m = \frac{y_A - y_B}{x_A - x_B} = \frac{19 - 11}{17\frac{1}{2} - 8} = \frac{8}{9\frac{1}{2}} = \frac{16}{19}$$

and the value of c can be found using this value of m along with the coordinates of B i.e. $11 = \frac{16}{19} \times 8 + c$ $\therefore c = 4\frac{5}{19}$

$\therefore$ the equation of the regression line is

$$y = \frac{16}{19}x + 4\frac{5}{19}$$

When x = 12, $y = \frac{16}{19} \times 12 + 4\frac{5}{19} = 14\frac{7}{19}$

When x = 22, $y = \frac{16}{19} \times 22 + 4\frac{5}{19} = 22\frac{15}{19}$

EXERCISE B

1. Find the equation of the regression line which passes through:
a) (7,8) and (4,2)   *b)* (2,5) and (5,2)
c) (5,6) and (21,14)   *d)* (2,3) and (8,7)

2. The grid shown below shows two regression lines. Find the equation of a) AB b) CD.

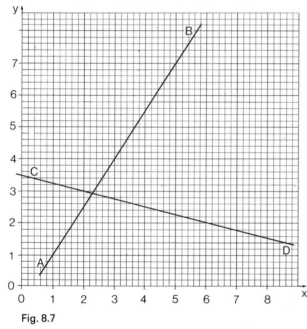

Fig. 8.7

3. Draw the scatter diagram corresponding to the following data.

x	3	7	11	13	20	25	30	37
y	6	9	10	12	14	18	21	23

x	86	90	93	94	95	96	98	100	100	105	110
y	120	104	100	80	74	47	98	25	20	50	10

On the diagram draw the best fitting line and find its equation. Use the equation to estimate the value of y which corresponds to x = 34.

4. Draw the scatter diagram and the best fitting line for the following set of measures of x and y.

x	2.2	2.8	3.4	4.0	4.2	5.4	6.0
y	7.4	6.4	6.0	4.5	3.8	1.8	1.6

Find the equation of the best fitting line and use that equation to find 1) the value of y when x = 3, 2) the value of y when x = 5.

5. The marks obtained by 8 students in a formal examination and in an objective test in mathematics are tabulated below. The marks listed under the symbol x are those for the formal test and under y are the marks in the objective test.

x	5	10	14	15	24	27	32	33
y	10	11	12	15	15	17	20	20

Illustrate these measures on a scatter diagram, draw the regression line of y on x and find its equation.

6. Draw a scatter diagram to illustrate the following measures. On the diagram draw the regression line of y on x and find its equation.

x	5	9	15	15	23	23	30	32
y	10	2	10	18	16	22	26	32

7. Draw a scatter diagram to illustrate the distribution below.

On the scatter diagram draw and find the equation of a) the regression line of y on x b) the regression line of x on y.

90

8. A body is dropped from a helicopter and its height above ground level is estimated at various times after release. The following table gives the results of these observations where y measures the height above ground level in metres and t measures the time after release in seconds.

t	2	3	4	5	6	7	8
y	380	360	320	270	220	160	80

Since it is known that the height should be given by a formula of the type $y = mt^2 + c$ draw up a table showing the mapping of y against t^2 and illustrate this mapping on a scatter diagram. Draw the best fitting line for this diagram and use it to find the values of m and c. Assuming that the regression line can be extended, estimate 1) the height from which the body was dropped i.e. the value when $t = 0$, 2) the time taken by the body to reach the ground i.e. the value of t when $y = 0$.

9. The cost of producing lace decreases as the amount of lace produced increases and some measures relating to this fact are illustrated in the following table where x is the length of lace produced in hundreds of metres and y is the cost of the lace in pence per metre.

x	$\frac{1}{4}$	$\frac{1}{3}$	$\frac{1}{2}$	2.	3	6
y	29	22	15	7.5	5.5	5

Since it is known that x and y are connected by an equation of the form $y = \frac{m}{x} + c$ draw

up a table showing the corresponding values of $\frac{1}{x}$ and y and plot these results on a scatter diagram. Draw the best fitting line for this diagram and use it to find the values of m and c. Assuming that the regression line can be extended, estimate a) the cost per metre of lace if 1000 metres are produced, b) the number of metres of lace that must be produced in order that the cost per metre is 4.5 pence.

Correlation

While regression measures the line which best fits a certain set of measures it takes no account of how closely these measures approximate to that line. In the study of correlation the interest is in measuring the degree by which two distributions match each other. The type of measure that is used for this purpose is called the **coefficient of correlation** and it is assessed on a scale which runs from +1 through zero to −1. A coefficient of correlation of +1 means that the two distributions match each other perfectly and this would correspond to a scatter diagram like that shown in Figure 8.8 below where all of the points plotted lie along the leading diagonal of the grid. A coefficient of correlation of −1 would correspond to a pair of distributions where the measures are in completely opposite order i.e. the first of the one distribution is last in the other, the second in the first distribution is second last in the other distribution and so on. This type of situation would correspond to a scatter diagram like that shown in Figure 8.9 below where all of

Fig. 8.8

Fig. 8.9

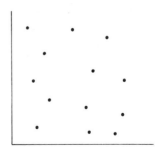

Fig. 8.10

the points plotted lie along the other diagonal of the grid. A zero coefficient of correlation would correspond to a pair of distributions which were completely unconnected and would have a scatter diagram like Figure 8.10 where the entries are scattered at random over the grid. Intermediate values of the coefficient of correlation correspond to distributions whose scatter diagrams are intermediate between these extreme cases.

There are fundamentally two methods of assessing a coefficient of correlation, the one depends on the actual measures in the two separate distributions being considered and is called the Product-Moment Coefficient of Correlation and the other which is called the Coefficient of Rank Correlation depends only on the order of merit of the measures in the distributions. The first of these methods involves a considerable amount of heavy calculation and thus in this study only the second of these methods of arriving at a measure of correlation will be used.

Coefficient of rank correlation

The theory behind this measure is beyond the scope of the present study and thus only the procedure will now be discussed. Each distribution must first be put into an order of merit and traditionally this is taken in a descending order. Each item being considered has now two ranks allocated to it and the

difference between these two ranks can now be found and if the symbol d is used to represent this difference then:

Coefficient of Rank Correlation

$$= R = 1 - \frac{6 \sum d^2}{n(n^2 - 1)}$$

where n is the number of items in the distributions.

If two or more measures in one distribution are equal it is convenient, though not mathematically justifiable, to allocate to them a rank which is the average of the ranks which they would have occupied if they had been different e.g. if the 3rd and 4th measures in a distribution are equal they would both be allocated the rank $3\frac{1}{2}$ or if the 5th, 6th and 7th are equal they would all be allocated the rank 6.

Even measures which cannot be measured numerically or exactly can be used in this context. All that is required is that the items can be put into an order of merit in respect of the quality being measured e.g. depth of colour or age of trees or rocks.

Example
Consider the distributions of reading speeds and intelligence quotients quoted in the earlier section of this chapter on scatter diagrams. The calculation of the coefficient of rank correlation can be tabulated as follows.

Child	Reading Speed	Intelligence Quotient	Rank in Reading Speed	Rank in Intelligence Quotient	d	d²
A	120	80	11	12	−1	1
B	140	90	$8\frac{1}{2}$	$10\frac{1}{2}$	−2	4
C	100	90	12	$10\frac{1}{2}$	$+1\frac{1}{2}$	$2\frac{1}{4}$
D	170	100	7	$8\frac{1}{2}$	$-1\frac{1}{2}$	$2\frac{1}{4}$
E	130	100	10	$8\frac{1}{2}$	$+1\frac{1}{2}$	$2\frac{1}{4}$
F	190	105	5	7	−2	4
G	220	110	3	$5\frac{1}{2}$	$-2\frac{1}{2}$	$6\frac{1}{4}$
H	140	110	$8\frac{1}{2}$	$5\frac{1}{2}$	+3	9
I	180	115	6	4	+2	4
J	240	120	2	3	−1	1
K	200	125	4	2	+2	4
L	270	130	1	1	0	0

$$\sum d^2 = 40$$

$$\text{Coefficient of Rank Correlation} = R = 1 - \frac{6 \sum d^2}{n(n^2 - 1)}$$

$$= 1 - \frac{6 \times 40}{12 \times 143}$$

$$= 1 - 0.14$$

$$= \underline{0.86}$$

A high positive coefficient of correlation such as the above indicates that it is likely that the measures are closely linked but it must be remembered that no conclusive results can be obtained from such a small sample.

EXERCISE C

1. Calculate the coefficient of rank correlation for the following distributions of x and y (ref. covered example in section on drawing of regression lines)

x	10	15	23	30	35	45
y	11	19	18	27	29	33

2. Calculate the coefficient of rank correlation of the distributions listed in Question 3 of Exercise B of this Chapter.

3. Calculate the coefficient of rank correlation of the Distribution listed in Question 4 of Exercise B of this Chapter.

4. Calculate the coefficient of rank correlation of the distributions listed in Question 5 of Exercise B of this Chapter.

5. Calculate the coefficient of rank correlation of the distributions listed in Question 6 of Exercise B of this Chapter.

6. Calculate the coefficient of rank correlation of the distributions listed in Question 7 of Exercise B of this Chapter.

7. Six candidates sat a written examination and were also interviewed for the award of a bursary. The written examination was marked out of 100 and the interview was graded A, or B, or C, or D, or E in descending order of merit. The results were as follows:

Candidate	A	B	C	D	E	F
Written Examination	80	78	60	55	53	48
Interview Result	C	A	B	B	D	C

Find the coefficient of rank correlation for the two sets of measures.

8. The marks in Mathematics and French for a group of 10 children are listed below. Calculate the coefficient of rank correlation for these measures.

Pupil	A	B	C	D	E	F	G	H	I	J
Mathematics	18	32	44	47	56	71	76	84	86	93
French	40	68	53	35	84	75	58	26	48	77

9. The heights, weights and shoe sizes of a group of 15 senior boys and 15 senior girls, in a school, are shown below.

Girls Height (cm)	Weight (kg)	Shoe size
154	66	$5\frac{1}{2}$
170	57	$6\frac{1}{2}$
173	80	8
163	54	$4\frac{1}{2}$
160	53	5
164	54	5
163	53	6
160	62	$4\frac{1}{2}$
160	45	$4\frac{1}{2}$
163	53	5
168	60	7
162	55	$3\frac{1}{2}$
169	60	7
165	57	4
163	57	6

Boys Height (cm)	Weight (kg)	Shoe size
181	71	$9\frac{1}{2}$
173	70	8
184	69	$8\frac{1}{2}$
183	57	7
163	57	7
184	79	9
170	62	$8\frac{1}{2}$
180	73	9
165	54	$7\frac{1}{2}$
178	70	11
168	54	7
178	60	8
180	68	9
180	65	8
173	70	$9\frac{1}{2}$

Calculate the coefficient of rank correlation for:
  a) height against weight of girls
  b) height against weight of boys
  c) height against shoe size of girls
  d) height against shoe size of boys.

10. Find the order of merit of the teams in the first division of the English Football League at the end of each of the last two years. (Ignore the teams which were relegated and promoted at the end of the first of these two years.) Find the coefficient of rank correlation for these two orders of merit. Repeat this exercise using the first division of the Scottish League.

9 PROBABILITY

What does probability mean to you?
The name rather suggests a vague likelihood of something happening. There is nothing vague however, about probability. Mathematically, probability is as precise as any other subject. We can calculate very exactly the probability or chance of a certain event happening when we know the circumstances surrounding it.

If we toss a **fair** (or honest) coin it will land head or tail upwards. There is a '50–50' chance of either. In other words, there is a probability of $\frac{1}{2}$ that a head shows and the same for a tail.

If we are interested in obtaining a head, then we say a head is a *favourable* outcome. The total number of possible outcomes is 2 (a head or a tail). The probability of obtaining a head is 1 out of 2 or $\frac{1}{2}$.

Let us consider rolling a die. The die could show a 1, 2, 3, 4, 5, or 6, i.e. there are 6 possible outcomes when a die is thrown. If we are interested in obtaining a 3, then there is a chance of 1 in 6 or $\frac{1}{6}$ of getting a 3; there is also a chance of $\frac{1}{6}$ of getting a 1 or 2 or 4 or 5 or 6.

The probability of a 3 showing

$$= \frac{\text{number of favourable outcomes}}{\text{number of possible outcomes}}$$
$$= \frac{1}{6}$$

What if we wanted to know the probability of a number divisible by 2 showing when a die is rolled? Again the total possible number of outcomes is 6. Now we must calculate the number of favourable outcomes i.e. how many of the numbers are divisible by 2. The numbers divisible by 2 are 2, 4 and 6. Thus the probability of a number divisible by 2 showing

$$= \frac{\text{number of favourable outcomes}}{\text{number of possible outcomes}}$$
$$= \frac{3}{6}$$
$$= \frac{1}{2}$$

So far, the probabilities we have considered have been purely theoretical probabilities. Let us now look at some actual results obtained experimentally.

EXPERIMENT 1   Tossing a coin

You should all have a coin to toss and a book and pencil to record your results.

Part 1
Copy this table into your book.

Number of Tosses = 50

	Tally	Total
Heads		
Tails		

You must make sure before you begin that you are tossing the coin *properly*. It must

95

spin round in the air before it falls. Now toss the coin 50 times and record the number of heads and tails obtained.

Tally your score as you did for frequency distributions. Finally total up your scores.

Let us compare the experimental results with what could be expected theoretically.

The probability of a head in one toss

$$= \tfrac{1}{2}$$

The expected number of heads in 50 tosses

$$= \tfrac{1}{2} \text{ of } 50$$

$$= 25$$

Similarly, the expected number of tails in 50 tosses is 25. Now, compare your experimental results with the theoretical results in this table.

Number of tosses = 50

	Experimental	Theoretical
Heads		25
Tails		25

Part 2

Repeat your experiment, tossing the coin 100 times. Again use a table to record the results. Calculate the theoretical results and compare them with the experimental results as in **Part 1**.

Part 3

Let us now collect together the number of heads and tails for the whole class for the 100 tosses.

Copy this table into your book. Each member of the class should read out his results in turn and everyone should record all the results. As before, calculate the theoretical results and compare them in the table.

Total number of tosses =

	Experimental		Theoretical	
	Heads	Tails	Heads	Tails
Total				

It should have become obvious that the greater the number of tosses of a fair coin we consider, the nearer the experimental results come proportionally to the theoretical.

EXPERIMENT 2   Throwing a die

A similar experiment may be done using a die. Each member of the class should have a fair (honest) die and a shaker.

Part 1

Copy this table into your book.

Number of throws = 60

	Tally	Total
1		
2		
3		
4		
5		
6		

Roll the die 60 times and record as before the number of times 1, 2, 3, 4, 5, 6 turn up.

Now, compare the experimental and theoretical results in a table.

The probability of a 1 in 1 throw

$$= \frac{\text{favourable}}{\text{possible}}$$

$$= \frac{1}{6}$$

The expected number of 1s in 60 throws

$$= \frac{1}{6} \text{ of } 60$$

$$= 10$$

Similarly, the expected number of 2s, 3s etc.

$$= 10$$

Number of throws = 60

	Experimental	*Theoretical*
1		10
2		10
3		10
4		10
5		10
6		10

Part 2
Repeat the experiment, rolling the die 120 times and recording your scores in a table as before. Calculate the theoretical results and compare them, in a table, with the experimental results.

Part 3
As in Experiment 1, collect together the total numbers of 1s, 2s, 3s etc. that the whole class had for Part 2. Again calculate the expected theoretical values. Complete this table.

Total Number of Throws =

1	2	3	4	5	6

Experimental totals

Theoretical totals

It should be clear, again, that the experimental results come closer proportionally to the theoretical results as we use a larger number of throws of the die.

Further experiments are quite easily devised for comparing experimental and theoretical results.

Suggestions
1. A solid regular triangular pyramid (tetrahedron) with the numbers 1, 2, 3, 4 painted on its faces may be rolled in a similar way to a die.
2. A spinner, as shown, can be made easily (but care must be taken). The disc may be marked off into any number of equal segments and either coloured or numbered as wished. The arrow should be able to spin round freely.

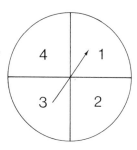

With a spinner of this type, a great many different experiments may be devised.

3. Various experiments may be conducted using books, e.g. opening a book at random and counting the number of times the tens digit is odd (or even) or a multiple of 2 or 3 etc.

4. On the market is a small roulette wheel that people can use at home. If some member of the class could borrow one of these, many interesting probability experiments could be tried.

Bias

You may have noticed that in the experiments with the coin and the die, the word 'honest' or 'fair' was used.

An honest coin is one which is equally likely to show a head or a tail.

An honest die is one in which each number has an equal chance of turning up.

A coin or die which is not honest is said to be **biased** and it will give biased results.

A two-headed coin would obviously be biased since there is no chance at all of getting a tail.

Dice that are used for gambling can be 'loaded' or weighted so that they tend to land a certain way up.

In a later chapter you will learn how you can test the honesty of a coin or a die, if there are any doubts about it.

In the suggested experiments, you must be wary of introducing any bias. The spinner, for example, would have to be made very carefully. The material of the platform must be of uniform thickness and the circle perfectly round. The segments also must be exactly equal. Another place where great care is required, is in the spinning arrow. It must be able to spin round quite freely.

In the experiment with the book, the book you choose to use, should not be one that has been habitually opened at certain pages or you will find that it opens at these pages more or less automatically. So, examine the book first to ensure that this sort of thing is not likely to happen.

Calculating probabilities

One of the difficulties in calculating probabilities is in working out the number of favourable outcomes and the number of possible outcomes. Each example must be considered carefully. Once we know these two quantities, we can very easily calculate the probability using this formula.

Probability of an event happening
$$= \frac{\text{Number of favourable outcomes}}{\text{Number of possible outcomes}}$$

If an event is certain to occur then the number of favourable outcomes is equal to the number of possible outcomes thus:

The probability of an event which is certain to occur $= 1$.

If an event is certain not to happen then the number of favourable outcomes is 0 and thus:

The probability of an event which is certain not to occur $= 0$.

If the probability that an event occurs is p and the probability that the event does not occur is q then:

$$p+q = 1 \text{ or } q = 1-p.$$

Example 1
A bag contains 50 marbles, 40 black ones and 10 red.
  a) What is the probability of drawing a red marble?
  b) What is the probability of drawing a black marble?
  c) If two black marbles are removed from the bag what is the chance now of drawing a black one?
  d) What is the chance of drawing a white marble?
  e) What is the probability of drawing either a red marble or a black one?

Solution
a) Number of favourable
   outcomes $= 10$
   Number of possible
   outcomes $= 50$

Probability of drawing a
red marble $= \dfrac{\text{favourable}}{\text{possible}}$

$$p\,(\text{red}) = \dfrac{10}{50}$$
$$= \dfrac{1}{5}$$

b) Number of favourable
outcomes $= 40$

Number of possible
outcomes $= 50$

Probability of drawing a
black marble $= \dfrac{\text{favourable}}{\text{possible}}$

$$p\,(\text{black}) = \dfrac{40}{50}$$
$$= \dfrac{4}{5}$$

c) Number of favourable
outcomes $= 38$

Number of possible
outcomes $= 48$

Probability of drawing a
black marble $= \dfrac{\text{favourable}}{\text{possible}}$

$$p\,(\text{black}) = \dfrac{38}{48}$$
$$= \dfrac{19}{24}$$

d) Number of favourable
outcomes $= 0$

Number of possible
outcomes $= 50$

Probability of drawing a
white marble $= \dfrac{\text{favourable}}{\text{possible}}$

$$p\,(\text{white}) = \dfrac{0}{50}$$
$$= 0$$

(i.e. there is no possible
chance of drawing a white
marble)

e) Number of favourable
outcomes $= 50$

Number of possible
outcomes $= 50$

Probability of drawing a
red or black marble $= \dfrac{\text{favourable}}{\text{possible}}$

$$p\,(\text{red or black}) = \dfrac{50}{50}$$
$$= 1$$

(i.e. it is a certainty that
either a red or a black
marble is drawn).

Example 2

Two dice thrown at the same time. What is
the probability that the sum of the two
numbers shown is six?

Solution

In this example, the difficulty lies in working
out all the possible outcomes. The simplest
way to do this, is to write down, in order, all
the pairs of numbers which could possibly
show on the dice. The first number refers to
the first die and the second to the other die.

1, 1	2, 1	3, 1	4, 1	5, 1	6, 1
1, 2	2, 2	3, 2	4, 2	5, 2	6, 2
1, 3	2, 3	3, 3	4, 3	5, 3	6, 3
1, 4	2, 4	3, 4	4, 4	5, 4	6, 4
1, 5	2, 5	3, 5	4, 5	5, 5	6, 5
1, 6	2, 6	3, 6	4, 6	5, 6	6, 6

Thus we see that there are 36 possible
outcomes, and 5 favourable outcomes.

Probability that the sum of
the numbers is 6 $= \dfrac{\text{favourable}}{\text{possible}}$

$$= \dfrac{5}{36}$$

1. A bag contains 30 marbles, 25 green ones and 5 red. What is the chance of picking out a green marble?

2. If a letter is taken at random from the word CHRYSANTHEMUM, what is the probability that
    a) it is a vowel?
    b) it is 'M'?

3. On throwing a die, what is the probability of turning up
    a) a 6?
    b) a number less than 3?
    c) a number more than 6?
    d) an odd number?

4. If we take a standard pack of 52 playing cards, what is the chance of drawing
    a) an ace?
    b) the ace of clubs?
    c) a heart?
    d) a king, queen or jack?
    e) a joker?

5. A black card is removed from a pack of cards. What is the probability of drawing
    a) a black card?
    b) a red queen?
    c) a king?

6. If a number is chosen at random from the numbers 1 to 30 inclusive, what is the chance that a prime number is picked?

7. A box of 2 dozen pencils contains 6 with broken points. What is the probability of picking out one which has not got a broken point?

8. In a car park, there are 100 vehicles, 85 of them being cars, 10 lorries and 5 buses. If they are all equally likely to leave, what is the probability of
    a) a bus leaving first?
    b) a car leaving second? (If a bus left first).

9. A box contains 50 coloured pencils, 20 red, 15 blue and the rest green. What is the chance of drawing
    a) a red pencil?
    b) a blue or green pencil?

c) a yellow pencil?
d) a red, blue or green pencil?

10. Two dice are thrown. What are the probabilities that the total score is
    a) 5?
    b) 1?
    c) 10?
    d) 14?
    e) less than 13?

11. Two pennies are tossed together. What is the probability of two heads showing?

12. If three pennies are tossed together, what is the chance of
    a) three heads showing?
    b) less than 2 tails showing?
    c) more than 2 tails showing?

13. A coin and a die are thrown together. What is the probability of
    a) a head and a 3 showing?
    b) a tail and a 4 showing?
    c) a tail and a 7 showing?

14. A solid regular triangular pyramid (tetrahedron) has the numbers 1, 2, 3 and 4 marked on its 4 faces. If it is thrown like a die, what is the probability that
    a) the 1– face lands downwards?
    b) the sum of the three faces showing is an odd number?

15. If 2 tetrahedrons like the one in the last question are thrown together, what is the chance that
    a) the sum of the numbers face down is odd? .
    b) the sum of the numbers face down is a prime number?

16. If a tetrahedron and a coin are thrown together, what is the probability of getting
    a) an odd number and a head showing?
    b) a multiple of 2 and a tail?

17. A die and 2 coins are tossed together. What is the chance of obtaining
    a) 2 heads and an even number?
    b) 1 head, 1 tail and a number divisible by 3?
    c) 2 tails and a number less than 3?
    d) 1 head, 1 tail and a number greater than 3?

18. A pair of dice and a penny are thrown together. Find the probability of
    a) a head and a total score of 7 showing.
    b) a tail and a total score greater than 10 showing.
    c) a tail and a total score less than 13 showing.

19. Four coins are tossed together. What is the probability of obtaining 4 heads?

20. Considering question 12 and question 19 could you calculate the probability of getting 5 heads when 5 coins are tossed together, without making an array of the results?

Independent events

When the probabilities of certain events occurring are quite unconnected to one another, these events are said to be independent events. For example, if a coin and a die are tossed at the same time, the probabilities of getting a head with the coin and a 6 with the die are quite independent of each other.

The probability of a head and a 6 at the same time = $p(H) \times p(6)$

    i.e. $p(H) \times p(6) = \frac{1}{2} \times \frac{1}{6} = \frac{1}{12}$

This can be checked by listing an array of the possible results.
    H,1 H,2 H,3 H,4 H,5 H,6
    T,1 T,2 T,3 T,4 T,5 T,6

    $p(H \text{ and } 6) = \frac{1}{12}$

In general, if A and B are independent events, then
$$p(A \text{ and } B) = p(A) \times p(B)$$

Mutually exclusive events

If in an experiment there are only two possible outcomes, A and B, then if A occurs, B cannot possibly occur. These outcomes are said to be mutually exclusive and $p(A) + p(B) = 1$.

For example, if we pick out marbles from a bag containing 5 white and 15 red marbles,

$p(\text{white}) = \frac{5}{20} = \frac{1}{4}$ and $p(\text{red}) = \frac{15}{20} = \frac{3}{4}$

$p(\text{white}) + p(\text{red}) = \frac{1}{4} + \frac{3}{4} = 1$

If in an experiment there are four possible outcomes, A, B, C or D, then
$$p(A) + p(B) + p(C) + p(D) = 1$$

For example, if we have another bag containing 5 white, 15 red, 20 blue and 10 green marbles, then

$$p(\text{white}) = \frac{5}{50} = \frac{1}{10}$$
$$p(\text{red}) \;\;\;= \frac{15}{50} = \frac{3}{10}$$
$$p(\text{blue}) \;\;= \frac{20}{50} = \frac{2}{5}$$
$$p(\text{green}) = \frac{10}{50} = \frac{1}{5}$$

$p(\text{white}) + p(\text{red}) + p(\text{blue}) + p(\text{green})$
$$= \frac{1}{10} + \frac{3}{10} + \frac{2}{5} + \frac{1}{5} = 1$$

From the above, it can be seen that the probability of getting either blue or green,
$$p(\text{blue or green}) = \frac{30}{50} = \frac{3}{5}$$
i.e. $p(\text{blue or green}) = p(\text{blue}) + p(\text{green})$.

Also, the probability of getting neither blue nor green,
    $p(\text{neither blue nor green}) = 1 - \frac{3}{5} = \frac{2}{5}$

Example 1
A tetrahedron with its sides numbered 1, 2, 3, 4 and a die are rolled together. What is the probability of getting
    a) a 4 on the tetrahedron and a 4 on the die?
    b) an odd number on the tetrahedron and an even number on the die?

Solution
(These are independent events.)
    a) (4 on tetrahedron and 4 on die)
        = p(4 on tetrahedron) × p(4 on die)
        = $\frac{1}{4} \times \frac{1}{6} = \frac{1}{24}$
    b) p(odd no. on tetrahedron and even no. on die)
        = p(odd no. on tetrahedron) × p(even no. on die)
        = $\frac{1}{2} \times \frac{1}{2} = \frac{1}{4}$

101

Example 2

C and D are two mutually exclusive events with probabilities $\frac{1}{5}$ and $\frac{1}{10}$ (there are obviously **more** than two possible outcomes). Calculate

a) the probability of either C or D occurring,

b) the probability of neither C nor D occurring.

Solution

a) $p(C \text{ or } D) = p(C) + p(D) = \frac{1}{5} + \frac{1}{10} = \frac{3}{10}$

b) $p(\text{not } C \text{ or } D) = 1 - \frac{3}{10} = \frac{7}{10}$

Example 3

In an experiment there are three possible outcomes, A, B or C. If $p(A) = \frac{1}{5}$ and $p(B) = \frac{1}{4}$, calculate

a) $p(C)$,

b) the probability that either A or C occurs.

Solution

a) $p(C) = 1 - [p(A) + p(B)]$

$= 1 - (\frac{1}{5} + \frac{1}{4}) = 1 - \frac{9}{20} = \frac{11}{20}$

b) $p(A \text{ or } C) = p(A) + p(C)$

$= \frac{1}{5} + \frac{11}{20} = \frac{15}{20} = \frac{3}{4}$

EXERCISE B

1. If a coin and a die are tossed together, calculate

a) the probability of getting a tail with the coin and an even number with the die,

b) the probability of a head with the coin and a number less than three on the die,

c) the probability of a head with the coin and a multiple of 3 on the die.

2. The sides of a die are marked with the numbers 2, 2, 4, 4, 6, 6. If the die is rolled twice what is the probability of getting

a) a 4 each time?

b) either a 2 or a 6 each time?

If the die is rolled three times, what is the probability of getting

c) a 2 each time?

d) either a 4 or a 6 each time?

3. There are two spinners, one marked into equal sections numbered 1, 2, 3, 4, 5 and the second one marked into equal sections A, B, C. Calculate the probability of getting

a) a 2 and a B,

b) a 5 and an A,

c) an even number and an A,

d) an odd number and either B or C.

4. In an experiment there are three possible outcomes, A, B and C. If $p(A) = \frac{1}{3}$ and $p(B) = \frac{1}{4}$, calculate

a) $p(C)$,

b) the probability that either A or C occurs.

5. P, Q, R and S are four mutually exclusive events with

$p(P) = \frac{1}{10}, p(Q) = \frac{1}{5}, p(R) = \frac{1}{4}, p(S) = \frac{1}{3}$

Calculate the probability of

a) either P or Q occurring,

b) either Q or S occurring,

c) neither R nor S occurring,

d) neither P nor S occurring.

6. A, B, C and D are the four possible outcomes of an experiment.

$p(A) = \frac{1}{7}, p(C) = \frac{1}{3}, p(D) = \frac{1}{2}$

Calculate

a) $p(B)$,

b) the probability of either A or D occurring,

c) the probability that neither B nor C occurs,

d) the probability that C or D occurs,

e) the probability that neither A nor C occurs.

7. What is the probability of throwing 6 tails in 6 tosses of a coin?

8. What is the probability of obtaining a total of two with two throws of a die?

9. There are three spinners, the first marked into four equal sections numbered 1, 2, 3 and 4, the second is marked into three equal sections coloured red, blue and yellow, and the third has five equal sections A, B, C, D and E. Calculating the probability of getting
    a) a 3, blue and D,
    b) an even number, red and A,
    c) a 1, red and E,
    d) an odd number, yellow and a vowel,
    e) a prime number, either red or blue and a B.

Calculating expected frequencies

The idea of calculating expected frequencies was introduced earlier in the chapter, in the experimental work.

If the probability of a head turning up when a coin is tossed is $\frac{1}{2}$, then when a coin is tossed 10 times you could expect 5 heads to show.

If the probability of obtaining a 6 when a die is thrown is $\frac{1}{6}$, then in 60 throws of a die you could expect $\frac{1}{6}$ of 60 or 10 sixes to show.

The expected frequency = probability of the event happening × number of trials.

Example
A die is thrown 50 times. What is the expected frequency of an even number turning up?

$$\text{Probability of an even number} = \frac{\text{favourable}}{\text{possible}}$$

$$p(\text{even}) = \frac{3}{6}$$

$$= \frac{1}{2}$$

The expected frequency of even numbers $= \frac{1}{2}$ of 50

$$= 25$$

EXERCISE C

1. If a coin is tossed 100 times, how many tails might you expect to turn up?

2. When a die is thrown 60 times, what is the expected frequency of
    a) a number less than 4 showing?
    b) a number less than 7 showing?
    c) a factor of 6 showing?

3. If a tetrahedron, with its faces marked 1 to 4, is thrown 80 times, how many times would you expect
    a) the two-face to be down?
    b) the sum of the faces showing to be more than 7?

4. Two tetrahedrons are thrown 160 times. How many times would you expect the sum of the 2 faces that land downward to be an odd number?

5. What is the expected frequency of two heads showing, if 2 coins are tossed together 80 times?

6. A pack of cards is shuffled and then a card is drawn. This card is replaced and the experiment is repeated another 149 times. Approximately how many times would you expect to pick
    a) a face card?
    b) a black card?

7. A town has 1250 children under the age of 10 years who have never had chickenpox. If it is known, from experience, that the probability of a child under 10 catching chickenpox is 0.16, how many of these children would you expect to take chickenpox?

8. In a certain town, 10% of the car drivers have some kind of accident, however slight, in a year's driving. If there are 3672 drivers in the town, what is the expected number of accidents in the coming year?

9. If the probability of having defective hearing is 0.08 in school children, how many pupils in a school of 1000 could be expected to have defective hearing?

103

10. Imagine the chance of a footballer breaking a leg during a football season is 0.15. If 60 players sign on with a club at the start of the season, how many of them are likely to break a leg during the season?

11. If the probability that a child has taken measles by the age of 12, is 0.6, how many of the 300 children aged 12 in a school could be expected not to have had measles?

EXERCISE D Miscellaneous

1. A firm produced the following breakdown of their employees and the occurrence of accidents during the past year (no employee had more than one accident during the year).

	Number of staff with		
	No accident	A minor accident	A serious accident
Production staff	356	43	3
Maintenance staff	24	5	1
Office staff	66	2	0

a) Calculate the total number of employees.
b) If a member of staff is chosen at random find the probability that he/she is
   1) in production staff and had a minor accident,
   2) in office staff and had a serious accident,
   3) in production or maintenance staff and had an accident of some nature.

2. When a certain coin is tossed the probability of getting a head is 0.6. For this coin find
   a) the probability of getting a tail,
   b) the probability of getting 2 heads if the coin is tossed twice,
   c) the probability of getting at least one head if the coin is tossed twice,
   d) the expected number of heads if the coin is tossed 60 times.

3. Explain why the following statements cannot all be true (P, Q and R are three mutually exclusive results of an experiment).
   The probability of P happening = 0.3
   The probability of Q happening = 0.4
   The probability of R happening = 0.5

4. Of the pupils in a class one-third of them use public transport to come to school.
   a) What is the probability that
      1) If two members are chosen at random both of them use public transport to come to school?
      2) If two members are chosen at random at most one of them uses public transport to come to school?
      3) If three members are chosen at random none of them uses public transport to come to school?
   b) On a particular day only 75% of the class are present. If a pupil is chosen at random find the probability that he/she was present that day and came to school by public transport.

5. It is estimated that three in ten people in Britain go abroad for their holiday.
   a) Calculate the probability that
      1) If three people are interviewed at random none of them went abroad on holiday last year.
      2) If three people are interviewed at random at least one of them went abroad on holiday last year.
   b) If 100 samples each of three people were taken, how often would we expect to find that all three of the people in the sample had gone abroad on holiday last year?

6. Two fair six-sided dice are thrown at the same time.
   a) Find the probability that the sum of the scores is (1) 2 (2) 3 (3) 4 (4) 5 (5) 6.
   b) If the two dice are thrown 100 times, calculate how often we would expect to get a total score of (1) 2 (2) 3 (3) 4 (4) 5 (5) 6 (6) 7 (7) 8 (8) 9 (9) 10 (10) 11 (11) 12.

10 THE BINOMIAL DISTRIBUTION

The word binomial contains the prefix *bi* meaning two. In the Binomial distribution we are concerned with *two* possible events occurring. If a coin is tossed, it will either come down heads or tails. If you go out in the street, the next person you meet will be either male or female. Next Tuesday will either be your birthday or not. In each case there are two possibilities.

If we decide which of the two possibilities is desirable, then there is a probability of its happening and a probability of its not happening. In Statistics, we normally say that the probability of a certain event happening is **p** and of its not happening is **q**.

Suppose we have a bag containing 100 balls, some red and some white. If there are 50 of each colour, we have an equal chance of drawing a white one or a red one. Thus the probability of obtaining a red ball is 50% or $\frac{1}{2}$ or 0.5, and the probability of obtaining a white one is the same. The two probabilities add up to 1.

Let us now suppose that in the bag there are 90 red balls and 10 white ones. The probability of drawing a red one is now $\frac{90}{100}$ or $\frac{9}{10}$ or 0.9, and the probability of failing to draw a red one is $\frac{10}{100}$ or $\frac{1}{10}$ or 0.1. Again the two probabilities add up to 1.

If p is the probability of drawing a red ball and q is the probability of not drawing a red ball, then in each of the above cases, we can say

$$p+q = 1$$

Consider another example. A school caretaker was clearing out old text books and he put a large number of English books and Geography books into the same sack, 30% of the books being English books and the rest Geography books.

Let p = probability of picking out an English book.

Let q = probability of picking out a Geography book.

In this case, $p = \frac{30}{100}$ or $\frac{3}{10}$ or 0.3

and, $\qquad q = \frac{70}{100}$ or $\frac{7}{10}$ or 0.7

Again $p+q = 1$ if we select one book. Suppose we want to pick out two books. This time there are four possible outcomes: 2 English books, 1 English book followed by 1 Geography book, 1 Geography book followed by 1 English book or 2 Geography books. The probabilities of the various possible outcomes are thus given by the formulae $p \times p$, $p \times q$, $q \times p$, $q \times q$ and we can further note that these terms can be further grouped as p^2, $2pq$, q^2 where p^2 is the probability of getting two English books, $2pq$ is the probability of getting one of each kind and q^2 is the probability of getting two Geography books. Since these are the only possible outcomes of our experiment their total must be 1 and thus we note that the terms of the formula $(p+q)^2 = p^2+2pq+q^2 = 1$ give the probabilities of the various outcomes of carrying out the experiment twice.

If we extend the above to the selection in turn of three books we have the position whereby each of the above results can be followed by the selection of an English or a Geography book. Thus the probability of selecting three English books is $p \times p \times p = p^3$. Secondly there are three ways in which we can select two English and one Geography books: the Geography one can be selected either first, second or third, and the probabilities of these various outcomes are $q \times p \times p$, $p \times q \times p$, $p \times p \times q$ and the total of these three terms, namely $3p^2q$ gives the probability of selecting two English and one Geography books. Similarly we can show that the probability of selecting one English and two Geography books is $3pq^2$ and of selecting three Geography books is q^3. The probabilities of the various possible outcomes of three trials are given by the expansion

$$(p+q)^3 = p^3 + 3p^2q + 3pq^2 + q^3 = 1.$$

This concept can be extended and thus the probabilities of the various outcomes of n trials of an experiment are given by the terms of the expansion of $(p+q)^n$.

Expansion of $(p+q)^n$

When dealing with low values of n it is reasonable to do the actual multiplications from first principles but for higher values it is essential that we look for a better technique.

If we consider the first few of the expansions of $(p+q)^n$ we obtain the following expansions:

$$(p+q)^1 = p+q$$
$$(p+q)^2 = p^2 + 2pq + q^2$$
$$(p+q)^3 = p^3 + 3p^2q + 3pq^2 + q^3$$
$$(p+q)^4 = p^4 + 4p^3q + 6p^2q^2 + 4pq^3 + q^4$$

A study of permutations and combinations may lead us to see that the coefficients of the terms in these expansions can be written in terms of the ^nC_r symbols, e.g.
$$(p+q)^4 = p^4 + {^4C_1}p^3q + {^4C_2}p^2q^2$$
$$+ {^4C_3}pq^3 + {^4C_4}q^4,$$

and this can lead us to the assumption that the general expansion is:

$$(p+q)^n = p^n + {^nC_1}p^{n-1}q + {^nC_2}p^{n-2}q^2$$
$$+ \ldots + {^nC_r}p^{n-r}q^r + \ldots {^nC_n}q^n.$$

We shall now proceed to prove the truth of this expansion by means of induction. Let us assume that the formula is true for some value of n, say n = m, i.e. we assume that

$$(p+q)^m = p^m + {^mC_1}p^{m-1}q + {^mC_2}p^{m-2}q^2$$
$$+ \ldots + {^mC_r}p^{m-r}q^r + \ldots {^mC_m}q^m.$$

then

$$(p+q)^{m+1} = (p+q)^m(p+q)$$
$$= (p^m + {^mC_1}p^{m-1}q + {^mC_2}p^{m-2}q^2 +$$
$$\ldots + {^mC_{r-1}}p^{m-r+1}q^{r-1} + {^mC_r}p^{m-r}q^r +$$
$$\ldots + {^mC_m}q^m)(p+q)$$

It is impossible for us to carry out this multiplication for the general case but we can consider individual terms of the product. Let us take as our general term the one which involves q^r. We shall obtain two terms in the product which involve q^r and these will be obtained from the product of $^mC_{r-1}p^{m-r+1}q^{r-1}$ and q and the product of $^mC_rp^{m-r}q^r$ and p. Thus the general term in the product is

$${^mC_{r-1}}p^{m-r+1}q^r + {^mC_r}p^{m-r+1}q^r$$
$$= ({^mC_{r-1}} + {^mC_r})p^{m-r+1}q^r$$
$$= {^{m+1}C_r}p^{m+1-r}q^r$$

(Note that it can be proved that $^mC_{r-1} + {^mC_r} = {^{m+1}C_r}$)

Applying this result to all of the terms of our products we obtain

$$(p+q)^{m+1} = p^{m+1} + {^{m+1}C_1}p^mq$$
$$+ {^{m+1}C_2}p^{m-1}q^2 + \ldots + {^{m+1}C_r}p^{m+1-r}q^r$$
$$+ \ldots + {^{m+1}C_{m+1}}q^{m+1}.$$

i.e. if the expansion is true for n = m then it is also true for n = m+1, but the expansion was true for n = 1 and by the above it must thus be true for n = 1+1 = 2. Since it is true for n = 2 by the above it must also be true for n = 2+1 = 3 and by continuing this argument we can obviously prove the rule true for all positive integer values of n.

Thus for all positive integer values of n we have that:

$$(p+q)^n = p^n + {}^nC_1p^{n-1}q + {}^nC_2p^{n-2}q^2$$
$$+ \ldots + {}^nC_rp^{n-r}q^r + \ldots + {}^nC_nq^n.$$

and in this expansion the term ${}^nC_rp^{n-r}q^r$ gives the probability of getting $n-r$ successes and r failures in n trials of an experiment.

Pascal's triangle

Since some students may find the algebra of the previous section to be beyond the level of their interest we can now look at the number pattern known as Pascal's Triangle and use it to obtain the coefficients of $(p+q)^n$. This pattern of numbers is built up and can be proved a valid method of obtaining the coefficients by the same methods as we used in the last section.

It will be seen that each term in the triangle is found by adding together the two terms in the line above which lie on either side of it.

Pascal's Triangle, as stated gives the coefficients of the terms in the expansion of $(p+q)^n$

$$(p+q)^1 = 1p + 1q$$
$$(p+q)^2 = 1p^2 + 2pq + 1q^2$$
$$(p+q)^3 = 1p^3 + 3p^2q + 3pq^2 + 1q^3$$
$$(p+q)^4 = 1p^4 + 4p^3q + 6p^2q^2 + 4pq^3 + 1q^4$$
etc.   etc.

The way in which Pascal's Triangle is used is shown in the following examples.

Example 1
In a certain collection of bulbs 80% are daffodils and the rest are narcissi. Calculate the probability of obtaining 0, 1, 2, 3, and 4 narcissi in a group of 4 bulbs.

Solution

$$p = \frac{4}{5} \doteqdot \text{daffodils}$$

$$q = \frac{1}{5} = \text{narcissi}$$

PASCAL'S TRIANGLE

Number of Trials (n)	Coefficients in the expansion of $(p+q)^n$	Number of Different Combinations
1	1 1	2
2	1 2 1	3
3	1 3 3 1	4
4	1 4 6 4 1	5
5	1 5 10 10 5 1	6
6	1 6 15 20 15 6 1	7
7	1 7 21 35 35 21 7 1	8
8	etc.	9
etc.		etc.

It may already have become obvious that the number of different combinations is always 1 greater than the number of trials of the experiment. E.g. if we wish to pick out 6 books together, there are 7 possible different combinations of English and Geography books. (Work out the combinations and check for yourself.)

With 4 bulbs, the expansion is $(p+q)^4 = 1$
The coefficients from Pascal's Triangle are 1, 4, 6, 4, 1.

Thus,
$$(p+q)^4 = p^4 + 4p^3q + 6p^2q^2 + 4pq^3 + q^4$$

By substituting $p = \frac{4}{5}$ and $q = \frac{1}{5}$, we get

107

a) Probability of 0 narcissus $= p^4 = \frac{4}{5} \times \frac{4}{5} \times \frac{4}{5} \times \frac{4}{5} = \frac{256}{625}$

b)   ,,   ,, 1 ,, $= 4p^3q = \frac{4}{1} \times \frac{4}{5} \times \frac{4}{5} \times \frac{4}{5} \times \frac{1}{5} = \frac{256}{625}$

c)   ,,   ,, 2 ,, $= 6p^2q^2 = \frac{6}{1} \times \frac{4}{5} \times \frac{4}{5} \times \frac{1}{5} \times \frac{1}{5} = \frac{96}{625}$

d)   ,,   ,, 3 ,, $= 4pq^3 = \frac{4}{1} \times \frac{4}{5} \times \frac{1}{5} \times \frac{1}{5} \times \frac{1}{5} = \frac{16}{625}$

e)   ,,   ,, 4 ,, $= q^4 = \frac{1}{5} \times \frac{1}{5} \times \frac{1}{5} \times \frac{1}{5} = \frac{1}{625}$

(Check: $(p+q)^4$ must equal 1

$$\frac{256}{625} + \frac{256}{625} + \frac{96}{625} + \frac{16}{625} + \frac{1}{625} = \frac{625}{625} = 1$$

Example 2
If 6 pennies are tossed together, what is the probability of 6 tails occurring?

Solution

$$p = \frac{1}{2} = \text{heads}$$

$$q = \frac{1}{2} = \text{tails}$$

With 6 pennies, the expansion is
$$(p+q)^6 = 1$$

From Pascal's Triangle, the relevant coefficients are 1, 6, 15, 20, 15, 6, 1.

Thus
$(p+q)^6 = p^6 + 6p^5q + 15p^4q^2$
$\qquad\qquad + 20p^3q^3 + 15p^2q^4 + 6pq^5 + q^6$

q^6 represents 6 tails.

Probability of 6 tails $= q^6 = (\frac{1}{2})^6$

$$= \frac{1}{2} \times \frac{1}{2} \times \frac{1}{2} \times \frac{1}{2} \times \frac{1}{2} \times \frac{1}{2}$$

$$= \frac{1}{64}$$

Example 3
In a certain community $\frac{2}{3}$ of the residents are regular T.V. viewers. Fifty investigators are sent out to interview 5 people each.

a) Calculate the probability that 4 or more people out of the 5, are regular viewers.

b) How many investigators would you expect to report that 4 or more people only viewed T.V. occasionally?

Solution

$$p = \frac{2}{3} = \text{regular T.V. viewers}$$

$$q = \frac{1}{3} = \text{non-viewers or occasional viewers}$$

With 5 people, the expansion is $(p+q)^5 = 1$
From Pascal's Triangle, the coefficients are 1, 5, 10, 10, 5, 1.

Thus the expansion is
$(p+q)^5 = p^5 + 5p^4q + 10p^3q^2 + 10p^2q^3$
$\qquad\qquad\qquad\qquad + 5pq^4 + q^5$

a) Probability of 4 viewers $= 5p^4q = \dfrac{5}{1} \times \dfrac{2}{3} \times \dfrac{2}{3} \times \dfrac{2}{3} \times \dfrac{2}{3} \times \dfrac{1}{3} = \dfrac{80}{243}$

Probability of 5 viewers $= p^5 = \dfrac{2}{3} \times \dfrac{2}{3} \times \dfrac{2}{3} \times \dfrac{2}{3} \times \dfrac{2}{3} = \dfrac{32}{243}$

Probability of 4 or more viewers $= \dfrac{80}{243} + \dfrac{32}{243} = \dfrac{112}{243}$

b) Probability of 4 non-viewers $= 5pq^4 = \dfrac{5}{1} \times \dfrac{2}{3} \times \dfrac{1}{3} \times \dfrac{1}{3} \times \dfrac{1}{3} \times \dfrac{1}{3} = \dfrac{10}{243}$

Probability of 5 non-viewers $= q^5 = \dfrac{1}{3} \times \dfrac{1}{3} \times \dfrac{1}{3} \times \dfrac{1}{3} \times \dfrac{1}{3} = \dfrac{1}{243}$

Probability of 4 or more non-viewers $= \dfrac{10}{243} + \dfrac{1}{243} = \dfrac{11}{243}$

Expected no. of investigators $= \dfrac{11}{243} \times \dfrac{50}{1} = \dfrac{550}{243} = 2.3 = $ approx. 2.

Practical work

EXPERIMENT 1 *The frequency of heads when six coins are tossed together.*
For the experiment divide yourselves into groups of 6 pupils, each pupil having a coin to toss.
   Copy this table into your jotter.

No. of trials = 64

No. of heads	Tally	Total
0		
1		
2		
3		
4		
5		
6		

All 6 members of the group should toss their coins at the same time and count up the number of heads showing, recording this in the table. Repeat this process until 64 trials in all have been completed.

Let us compare these experimental frequencies with the expected frequencies. Calculate the expected frequencies of 0, 1, 2, 3, 4, 5 and 6 heads, when 6 coins are tossed simultaneously 64 times, by using the binomial expansion of $(p+q)^6$.
   Complete this table.

No. of heads	0	1	2	3	4	5	6
Experimental Frequency							
Theoretical Frequency							

Using this frequency table draw frequency polygons on the same graph for the experimental and theoretical frequencies of the number of heads showing when 6 coins are tossed together.
   Compare your graph with the graphs of the other groups.
   Make up another frequency table for experimental and theoretical results by totalling up the results of all the groups, and draw a graph as before. Compare this graph with your first one.
   What conclusions do you reach about the distributions and in particular about the symmetry of them?

EXPERIMENT 2 *The frequency of sixes when four dice are rolled together.*
This experiment may be done either by individuals or by groups of two or three pupils.

Copy this table into your book.

No. of trials = 72

No. of Sixes	Tally	Total
0		
1		
2		
3		
4		

Roll 4 dice together, 72 times in all, tallying the number of Sixes obtained each time and record the results in the table.

Now, calculate the expected frequencies of 0, 1, 2, 3, and 4 sixes when 4 dice are rolled together, 72 times, using the binomial expansion of $(p+q)^4$.

Fill in the results in the following table.

No. of Sixes	0	1	2	3	4
Experimental frequency					
Theoretical frequency					

Draw, on the same graph, frequency polygons for these experimental and theoretical frequencies of 0, 1, 2, 3, and 4 Sixes showing when 4 dice are rolled together. (In this graph consider your vertical scale carefully.)

Compare your graph with the graphs of the other members of the class.

As in Experiment 1, form another frequency table by adding up all the results for the class. Draw a graph for these frequencies. Compare this graph with your previous one.

What do you notice about the symmetry of this distribution? Consider the second graph of Experiment 1. What reasons can you give for points of similarity or difference?

110

FURTHER EXPERIMENTS

Various other experiments may be devised and carried out along the same lines as the above experiments.

Several pairs of dice may be rolled at the same time.

Several sets of dice may be rolled and a coin may be tossed simultaneously.

A tetrahedron with its faces numbered may be rolled in the same way as a die.

Several pairs of tetrahedrons may be thrown at the same time.
etc.

EXERCISE A

1. On the average, a marksman firing at a target hits the bull's-eye once in three shots. If he fires 6 times, what are the chances that he will hit the bull's-eye
   a) twice   *b)* 4 times *c)* not at all?

2. Another marksman, on average, hits the target 2 out of 3 times. In 4 shots what are his chances of hitting it 0, 1, 2, 3, or 4 times?

3. What is the probability of getting 4 heads in 8 tosses of a coin?

4. A shipment of oranges contains 10% bad ones. A random sample of 4 oranges is drawn from the shipment. Calculate the probability that the sample contains 0, 1, 2, 3 and 4 bad oranges.

5. In 10 throws of a coin what is the probability of obtaining 8 or more heads or tails?

6. In a certain large collection of plants, $\frac{2}{3}$ are single flowered varieties and the rest are double flowered. Calculate the probability of obtaining 0, 1, 2, 3, and 4 double flowered plants in a row of 4 plants.

7. Robert's chance of winning a set at tennis against David is $\frac{3}{4}$. Find his chance of winning at least 3 sets in a 5 set match assuming that all 5 sets are to be played.

8. An opinion poll finds that 3 out of 5 people are in favour of a certain proposal. What is the probability, that, if 3 people are taken at random, there will be a majority against the proposal?

9. In a certain town, the proportion of rainy to fine days in the month of June is 1 to 3. Assuming that each day is independent of the others, what is the chance that a week in June, in that town, will have
a) no wet days *b*) 3 wet days?

10. From a box containing 6 white balls and 4 black balls, 3 balls are drawn at random. Find the probability that 2 are white and 1 is black.

11. In a certain industry, the workmen have a 20% chance of contracting an occupational disease. What is the probability that out of 6 workmen, 4 or more will contract the disease?

12. What is the probability of throwing at least 3 sevens in 5 throws of a pair of dice?

13. A coin is tossed 6 times. What is the probability of getting
a) exactly 3 heads   *b*) at least 3 heads?

14. Seven dice are rolled. Calling a 5 or 6 a success, find the probability of having
a) exactly 4 successes
b) at most 4 successes

15. Assuming that the chance of a child being a boy or a girl is $\frac{1}{2}$, what is the probability that in a family of 6 there will be no fewer than 2 or more than 5 boys? What is the chance there will be 3 boys?

16. In a very large batch of screws, 5% are defective. What is the probability of finding at least 1 defective in samples of 5, 10, 15 and 20 respectively?

17. A manufacturer knows that an average 1 out of 10 of his products is defective. What is the probability that a random sample of 4 articles will contain
a) no defectives
b) exactly 1 defective
c) at least 2 defectives
d) no more than 3 defectives?

18. In a packet of flower seeds 40% are known to be pink flowering and the remainder white flowering. Calculate the probabilities of 0, 1, 2, 3, 4 or 5 pink flowers in a row of 5 plants. If 500 rows each of 5 plants are planted, approximately how many rows will contain
a) all pink flowers
b) all white flowers?

19. In a certain community, 2 out of every 3 houses has a telephone. If 50 investigators, each question 4 households, selected at random, how many may be expected to report that
a) only 1 household out of the four does not have a telephone
b) at least 2 out of the four households has a telephone?

20. One hundred interviewers are told to select 5 people at random and to question them concerning their drinking habits. Assuming that 3 out of 4 people drink alcohol at least occasionally, how many interviewers may be expected to find that
a) at least 2 people are total abstainers.
b) at least 3 people drink alcohol sometimes?

THE NORMAL DISTRIBUTION

Frequency distributions

Let us consider the following frequency distributions where the total frequency is fairly high.

Example 1

This distribution shows the number of children per family in the families of all first year pupils in a Secondary School in a country area. (This cannot be said to be quite typical of the families in the area as a whole since families with no children are not considered.)

Figure 11.1 shows the histogram for this distribution.

Number in family	1	2	3	4	5	6	7	8	9	10	11	12	13
Frequency	10	27	37	36	22	18	14	9	4	3	0	2	1

Fig. 11.1 Number of Children per Family

As you can see from the histogram, this is a rather 'lopsided' distribution. This kind of distribution is said to be **skewed.**

Example 2
The results recorded in a cycling proficiency test given by the police to school children in Wigtownshire are shown below. This was not a compulsory test, pupils could enter or not as they wished. Most of the children were aged about 10–12 years.

Marks	Frequency
35–39	1
40–44	0
45–49	0
50–54	1
55–59	4
60–64	3
65–69	6
70–74	16
75–79	22
80–84	75
85–89	113
90–94	68
95–99	5

Draw a histogram to illustrate this. Can you suggest any reasons for the shape of the distribution? Is this distribution skewed? What shape do you think it might have had if every child in the area aged 10–12 years had entered the test?

Example 3
These deaths (from all causes) were recorded one year in Scotland.

Age (years)	Frequency
0– 4	1619
5– 9	117
10–14	91
15–19	177
20–24	177
25–29	175
30–34	208

Age (years)	Frequency
35–39	392
40–44	682
45–49	960
50–54	1846
55–59	2951
60–64	4165
65–69	4181
70–74	4343
75–79	4082
80–84	3182
85–89	1814
90–94	523
95–99	73
100+	5

Without any further grouping, draw a histogram for this distribution and discuss the shape of it.

This shape of distribution is the usual one for deaths and it is an example of what is called a U type distribution.

Example 4

A ten sum test for speed and accuracy was given to all first year pupils in a country area and the results were as follows.

Number of sums correct	0	1	2	3	4	5	6	7	8	9	10
Frequency	5	11	21	35	49	52	51	28	20	13	3

Figure 11.2 (a) (on the next page) shows the histogram for this

This histogram is fairly symmetrical, as you might expect since it represents the complete range of abilities in this area.

The mean, median and mode are all 5.

In this chapter we are principally interested in distributions of this type, i.e. symmetrical about the mode and shaped like a bell. The standard distribution of this type is called a **Gaussian** or **normal distribution.**

113

Frequency

No. of sums correct

Fig. 11.2 Sum Test

Example 5

Some of the first year pupils who sat the above test, had already done this same test several months before when they were at the end of their primary education.

Number of sums correct	0	1	2	3	4	5	6	7	8	9	10
Frequency (Primary)	0	1	6	7	15	23	20	19	22	8	8
Frequency (Secondary)	3	2	9	16	20	23	18	18	11	7	2

On the same graph, draw frequency polygons for the two distributions, primary and secondary. Comment on the shape of the distributions and give reasons for the difference in the results from the one test to the next.

Note: The sums given in the test are shown opposite. The pupils sitting the test were allowed to write their names on the testpaper and then given *exactly* $3\frac{1}{2}$ minutes to answer it.

TEST PAPER

1. Add

55
346
107
432
——

2. Add

5038
967
2864
89
4692
——

3. Add

6875
428
5069
3786
53
8647
——

4. Subtract

7404
3856
——

5. Subtract

25983
 9087
——

6. Subtract

70136
49768
——

7. Multiply

276
× 48
——
——

8. Multiply

5397
× 76
——
——

9. Divide

9)8658

10. Divide

26)13962

114

Practical work

Collect some data of your own by undertaking some surveys. There are two principal types of surveys that may be conducted—those dealing with data about objects and those dealing with information about people.

Here are a few suggestions that you may care to follow up. A little thought and some help with apparatus from perhaps the science department, should enable them to be carried out fairly easily.

In each case a reasonably large number of observations should be taken and a frequency table constructed. Then the mean and standard deviation could be calculated and the results shown graphically as a histogram or frequency polygon.

Suggestions

1. An investigation could be conducted into the length of life of torch batteries. Here you would need to borrow some electrical equipment from the science department for the setting up of electrical circuits.

2. The height of bounce of a large number of tennis (or golf) balls could be measured. In this experiment, the balls must be dropped from a fixed height and the heights of the resulting bounce judged against a background board marked off in suitable measuring units. Obviously this work requires a team of pupils working together.

3. An investigation into the weight range of 'standard' eggs could be very interesting. You should be able to 'borrow' a large number of 'standard' eggs from the school kitchen or the cookery teacher and weigh them accurately with a science balance. To obtain enough weighings, the experiment could be spread over several weeks. Once you have formed a frequency distribution you will know the range of weights which covers the category of 'standard' eggs. Then you can contact an egg packing station and find out what the official range of weights is, and compare this with your own results.

4. A similar type of experiment may be done by very accurate weighing of packets of salt, custard powder, sugar etc. which have a stated weight on the packet.

5. A survey of the heights of a certain year group or age group in the school may be conducted quite readily (with some cooperation from the staff). All that is required is some kind of measuring instrument—at its simplest a metre stick tacked onto a wall. Here of course, boys and girls should be considered separately.

6. The same type of survey may be done by taking some other measurement of a group in the school, e.g. weights.

These last two surveys can be broadened in scope by contacting another school or several schools in various parts of the country and 'swopping' information with them. This would give several distributions for purposes of comparison. It would also be interesting to keep a record of things like heights and weights in a school over a number of years.

7. Many other surveys may be conducted in a school, e.g. sizes of shoes, the number of children per family, speed and accuracy sum tests, etc.

The normal distribution

In Example 4, at the beginning of the chapter, we saw a distribution symmetrical about the mode (and the mean and median) which approximated to a **normal distribution.** Figure 11.3 shows again the histogram of this distribution. This time a curve has been drawn through the tops of the columns. This is roughly the typical bell shaped curve of a normal distribution and is called the **normal curve.**

The normal curve

The algebraic study of this curve took place in the middle of the eighteenth century and is mainly attributed to De Moivre and Gauss. Their aim was to obtain the equation of a curve which corresponded to the

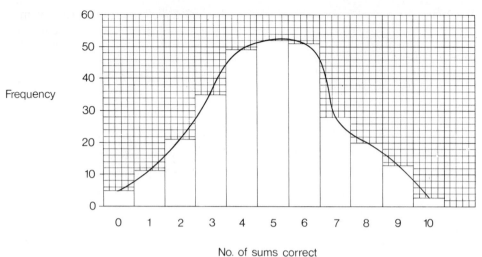

Frequency

No. of sums correct

Fig. 11.3 Sum Test

symmetrical bell shape. In the distribution the mean, median and mode coincide and the relative width of the 'bell' will depend on the spread of the distribution.

The details of this study are beyond the scope of this book but it is worth noting that the equation that is used in relation to the normal distribution curve is

$$y = \frac{N}{\sigma\sqrt{2\pi}}e^{\frac{-(x-\mu)^2}{2\sigma^2}}$$

where x is the continuous variable,
    N is the total number of observations
    μ is the mean of the distribution,
and σ is the standard deviation of the distribution.

The size and shape of the "bell" depend on the scales used. In b) the horizontal scale has been shortened and in c) the vertical scale has been lengthened.

Fig. 11.5 The Size and Shape of the "Bell" depend on the Scales Used

(a)

(b)

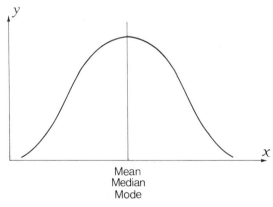

Mean
Median
Mode

Fig. 11.4 The Normal Curve

116

(c)

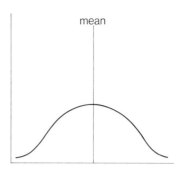

A distribution which is shaped like the normal curve is called a **normal distribution.** A great many distributions which occur naturally are normal distributions e.g. the weights of a certain age group of men, the number of tomatoes per plant, the number of peas in a pod. Your practical work should have produced normal distributions also.

The mean and standard deviation of the normal curve

The mean of a normal distribution lies in the middle, the curve being symmetrical. The standard deviation gives us the measure of the spread or dispersion of the observations.

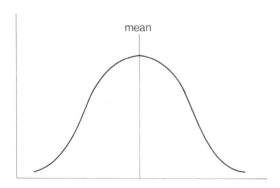

Fig. 11.6

Figure 11.7 shows 3 normal distributions with the same mean but differing standard deviations.

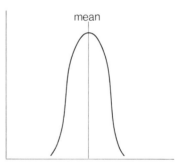

Fig. 11.7

It is clear that when examining any normal distribution we must take into account both the mean and the standard deviation.

In this way, we are able to compare normal distributions with one another, though in practice we compare all normal distributions with the special distribution which we have already called the Normal Curve.

117

Standard normal curve

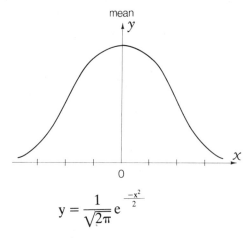

$$y = \frac{1}{\sqrt{2\pi}} e^{\frac{-x^2}{2}}$$

Fig. 11.8 Standard Normal Curve

Since the shape and position of each normal curve depends upon the mean and standard deviation of the distribution it will simplify our work if we standardise our data. Thus we obtain the **standard normal curve** which is a special version of the Normal Curve with the mean (μ) = 0, N = 1, and the standard deviation (σ) = 1. Further, the area between the curve and the x-axis is 1 unit and thus the area between the curve, the x-axis the ordinate x = a and the ordinate x = b is a measure of the probability that the variable x lies between x = a and x = b. These probabilities have all been accurately calculated and are usually tabulated in standard sets of tables. Some examples of these probabilities are illustrated below.

Fig. 11.9

Standardising scores

A point x on a normal curve with mean = μ and standard deviation = σ corresponds to the point Z on the standard normal curve when $Z = \dfrac{x - \mu}{\sigma}$

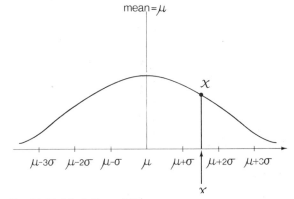

Fig. 11.10 (a) A Normal Curve

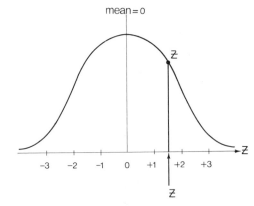

Fig. 11.10 (b) The Standard Normal Curve

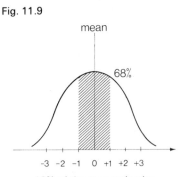

68% of the area under the curve lies with 1 Standard Deviation of the Mean.

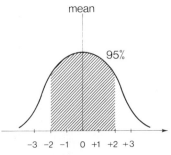

95% of the Area under the curve lies within 2 Standard Deviations of the Mean

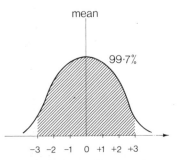

99.7% of the area under the curve lies within 3 Standard Deviations of the Mean

118

Thus we can calculate the corresponding standard score (Z) for any score (x) in a normal distribution by using the formula

$$Z \text{ (Standard Score)} = \frac{x - \mu}{\sigma}$$

(This process is called standardising the score.)

In this way, we can compare scores in different normal distributions by finding the standard score for each and seeing how they stand in relation to one another.

Example

In a statistics examination the mean mark was 78.0 and the standard deviation 8.0. In algebra, the mean was 73.0 and the standard deviation 7.6. A pupil scored 75 in statistics and 71 in algebra. In which examination was his relative standing higher?

Solution

Statistics: $\mu = 78.0$　　Algebra: $\mu = 73.0$

　　　　　$\sigma = 8.0$　　　　　　　$\sigma = 7.6$

　　　　　$x = 75$　　　　　　　　$x = 71$

$$Z = \frac{x - \mu}{\sigma} \qquad\qquad Z = \frac{x - \mu}{\sigma}$$

$$= \frac{75 - 78}{8} \qquad\qquad = \frac{71 - 73}{7.6}$$

$$= \frac{-3}{8} \qquad\qquad\quad = \frac{-2}{7.6}$$

$$= -0.375 \qquad\qquad = -0.263$$

i.e. the standard score 　i.e. the standard score
for statistics = −0.375　for algebra = −0.263
　　(0.375　　　　　　　　(0.263
　　below　　　　　　　　below
　　the　　　　　　　　　the
　　mean)　　　　　　　　mean)

Relatively speaking, the pupil did better in algebra than in statistics.

Note: It is of benefit in working this type of example to summarise the information given about each subject before going on to calculate the standard score.

EXERCISE A

1. Calculate the standard scores for a), b), c), d).

	a)	b)	c)	d)
x	72	45	63	71
μ	60	50	60	79
σ	9	5	11	5

2. A normal distribution has a mean $\mu = 120$ and a standard deviation $\sigma = 11$. What standard scores correspond to raw scores of a) 115, b) 134, c) 93?

3. Mary had a Geography mark of 63 in one examination where the average mark was 70 and the standard deviation 7. In the next examination, where the average was 71 she scored 65 (standard deviation being 5). Compare her standing in the two examinations.

4. John scored 75 in an arithmetic test where the mean mark was 70 and the standard deviation 6. His sister Jane scored 70 in her arithmetic test, the mean and standard deviation being 60 and 8 respectively. Which of two showed more ability in Arithmetic?

5. In an English examination the mean mark was 56 and the standard deviation was 8; in French the mean was 60 and standard deviation 14; in Mathematics, the mean and standard deviation were 65 and 10 respectively. A pupil scored 60 in English, 67 in French and 70 in Mathematics. Compare his standing in the three subjects.

6. Tom and Bob are friends who attend different schools. In the term examinations Tom had a mark of 70 in English and Bob had a mark of 80. The average mark in Tom's class was 65 with a standard deviation of 11. In Bob's class the average was 70 and the standard deviation was 10. Which of the boys had actually achieved the better result?

119

7. A pupil scored 50 in English and 62 in Mathematics in the November examination, the average marks being 55 and 60 respectively. In May, he scored 55 in English and 60 in Mathematics, the average mark being 58 for both subjects. The standard deviation for English was 7 in November and 5 in May; and for Mathematics was 5 in November and 12 in May.

    a) In which examination did he do better in English?

    b) In which examination did he do better in Mathematics?

    c) In which of all 4 examinations had he the best results?

    d) In which of all 4 examinations had he the poorest results?

8. In a normal distribution with mean 30 and standard deviation 3, what raw scores correspond to the following standard scores?

    a) 1.4   *b*) 2.5   *c*) -1.5 *d*) -0.68

9. In a normal distribution with mean 55 the raw score 60 corresponds to a standard score of 1.3. What is the standard deviation of the distribution (correct to the second decimal place)?

10. A normal distribution has a mean of 72, and a raw score of 60 corresponds to a standard score of -1.2. What is the standard deviation?

11. What is the mean of a normal distribution which has a standard deviation of 5 and in which a raw score of 30 corresponds to a standard score of 1.7?

12. In a normal distribution, the standard deviation is 2.3 and a score of 62 corresponds to a standard score of -0.5. What is the mean of the distribution?

13. What are the mean and the standard deviation of a normal distribution when raw scores of 45 and 60 correspond to standard scores of -1 and 2 respectively?

14. In a normal distribution scores of 70 and 50 correspond to standard scores of 0.5 and -1.5 respectively. Calculate the mean and standard deviation of the distribution.

120

Area under the standard normal curve

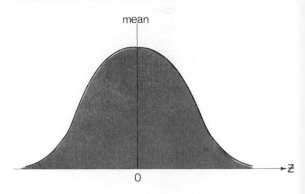

Fig. 11.11 (a) The total area under the Standard Normal Curve equals 1 (or 100%)

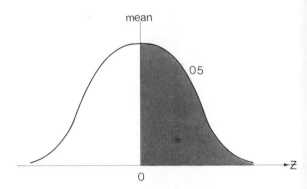

Fig. 11.11 (b) The area to the right of the mean (above the mean) equals 0.5 (or 50%). Similarly the area to the left of the mean (below the mean) equals 0.5 (or 50%)

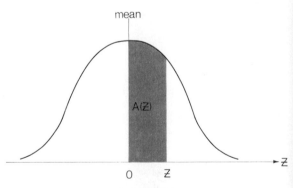

Fig. 11.11 (c) The shaded area under the curve A(Z) is the proportion of area between Z = 0 and any value Z

Tables have been compiled whereby we can read off the size of part of the area under the Standard Normal Curve provided we know the standard score (see Table 1 of Appendix).

Example 1
Find the area under the normal curve between $Z = 0$ and $Z = 1.2$.

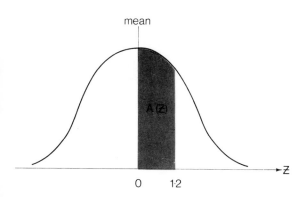

Fig. 11.12 (a) $Z = 1.2$ therefore A(Z) = 0.385

Example 2
Find the area under the normal curve to the right of $Z = 1.2$.

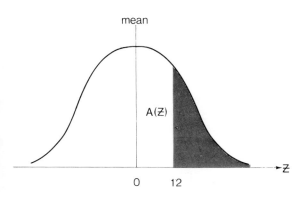

Fig. 11.12 (b) $Z = 1.2$ therefore A(Z) = 0.385
Required area = 0.5—0.385 = 0.115

Note: A (Z) is the area from the mean to Z. Since we require the area to the right of $Z = 1.2$, we must subtract A(Z) from the total area to the right of the mean which is 0.5.

Example 3
Find the area between $Z = 0$ and $Z = -1.2$

$$Z = -1.2$$
$$\therefore A(Z) = 0.385$$

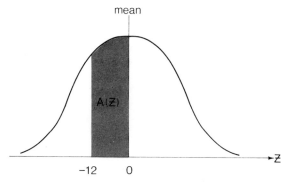

Fig. 11.13

(In the tables, only positive values of Z are given. Since the curve is symmetrical, the area for $Z = -1.2$ is the same as the area for $Z = +1.2$.)

Example 4
What is the area between $Z = 0.81$ and $Z = 1.94$?

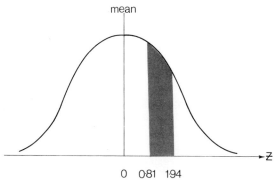

Fig. 11.14

$$Z_1 = 1.94$$
$$\therefore A(Z_1) = 0.474$$
$$Z_2 = 0.81$$
$$\therefore A(Z_2) = 0.291$$
$$\text{Required area} = 0.474 - 0.291$$
$$= 0.183$$

121

Example 5
Calculate the area from $Z = -0.46$ to $Z = 1.21$

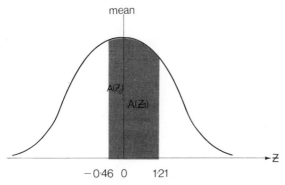

Fig. 11.15

$$Z_1 = 1.21$$
$$\therefore A(Z_1) = 0.387$$
$$Z_2 = -0.46$$
$$\therefore A(Z_2) = 0.177$$
$$\text{Required area} = 0.387 + 0.177$$
$$= 0.564$$

Example 6
A normal distribution has mean $\mu = 10$ and standard deviation $\sigma = 2$. Find the area under the curve from $x = 8$ to $x = 13$.

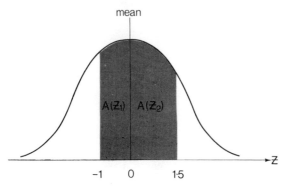

Fig. 11.16

First of all, we must standardise the scores.
$$x_1 = 8$$
$$Z_1 = \frac{x_1 - \mu}{\sigma} = \frac{8 - 10}{2} = \frac{-2}{2} = -1$$

122

$$x_2 = 13$$
$$Z_2 = \frac{x_2 - \mu}{\sigma} = \frac{13 - 10}{2} = \frac{3}{2} = 1.5$$
$$Z_1 = -1$$
$$\therefore A(Z_1) = 0.341$$
$$Z_2 = 1.5$$
$$\therefore A(Z_2) = 0.433$$
$$\therefore \text{Required area} = 0.341 + 0.433$$
$$= 0.774$$

Note: A diagram, as shown in the opposite column, for each example is of great benefit.

EXERCISE B
Find the area under the normal curve in each of the following cases.
1. from $Z = 0$ to $Z = 1.3$
2. from $Z = 0$ to $Z = 2.43$
3. from $Z = 0$ to $Z = -0.6$
4. from $Z = 0$ to $Z = -1.05$
5. from $Z = -0.6$ to $Z = 1.23$
6. from $Z = -1.78$ to $Z = 2.36$
7. from $Z = 0.4$ to $Z = 1.7$
8. from $Z = -0.62$ to $Z = -2.33$
9. from $Z = 0.1$ to $Z = 0.33$
10. from $Z = -1.6$ to $Z = -0.01$
11. to the right of $Z = 1.2$
12. to the right of $Z = 3.24$
13. to the left of $Z = -0.62$
14. to the left of $Z = -1.4$
15. to the right of $Z = -2.11$
16. to the right of $Z = -0.5$
17. to the left of $Z = 1.32$
18. to the left of $Z = 0.34$
19. for values of Z greater than 0.78
20. for values of Z greater than -0.78
21. for values of Z less than 1.30
22. for values of Z less than -0.88
23. A normal distribution has mean $\mu = 12$ and standard deviation $\sigma = 2$. Find the following areas under the normal curve.
   a) from $x = 10$ to $x = 13.5$
   b) from $x = 12.5$ to $x = 14$
   c) from $x = 8$ to $x = 9.6$

d) for values of x exceeding 14

e) for values of x less than 11

24. The following areas refer to areas under the standard normal curve. From the information given find the values of Ƶ (use the Tables of Area).

    a) The area between 0 and Ƶ is 0.379

    b) The area between 0 and Ƶ is 0.486

    c) The area to the left of Ƶ is 0.323

    d) The area to the right of Ƶ is 0.192

    e) The area to the right of Ƶ is 0.595

    f) The area to the left of Ƶ is 0.9

    g) The area between 0 and Ƶ is 0.493

    h) The area between 0 and Ƶ is 0.497

    i) The area between 0 and Ƶ is 0.490

Probability and area

Probability and the areas of histograms or distribution graphs are very closely connected. Let us consider this frequency distribution of the number of children per family in 50 families

No. of children	0	1	2	3	4	5	6	7
Frequency		3	6	12	15	8	3	2	1

The number of families out of the 50 with 3 children was 15. Thus the probability of choosing a family, at random, with 3 children in it is $\frac{15}{50}$.

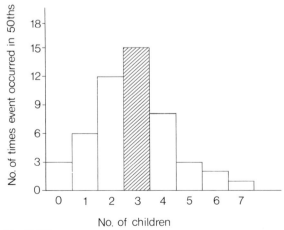

Fig. 11.17

The scale of the histogram may be chosen so that the total area enclosed by the columns is 1. To do so in this case, the vertical scale is in 50ths. Thus the area of the column representing the incidence of 3 children in a family is $\frac{15}{50}$.

So, the probability of picking out a family with 3 children is the same as the area of the column representing that incidence.

This same principle applies when we consider normal distribution curves. The area under the standard normal curve is 1. This means that the total probability is 1; and an area of 0.3 is the same as a probability of 0.3.

Example

In a normal distribution with mean 10 and standard deviation 2, what is the probability that x is greater than 12?

Solution

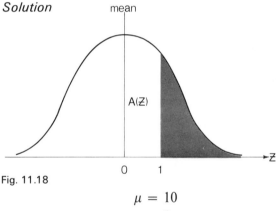

Fig. 11.18

$$\mu = 10$$
$$\sigma = 2$$
$$x = 12$$

(First we must standardise x)

$$\mathcal{Z} = \frac{x - \mu}{\sigma}$$
$$= \frac{12 - 10}{2}$$
$$= 1$$
$$A(\mathcal{Z}) = 0.341$$
$$\text{Required area} = 0.5 - 0.341$$
$$= 0.159$$

i.e. the probability of x being greater than 12 is 0.159.

EXERCISE C

1. Given that x is normally distributed with mean 50 and standard deviation 5, use the tables to calculate the probability that

a) x>60   b) x>55   c) x<43

d) x<40   e) 40<x<60

2. Assuming that the stature (x) of female university students is normally distributed with mean 160 cm and standard deviation 5 cm, calculate the probability that

a) x>152.5 cm

b) 150<x<165

The normal curve as a limit of a frequency distribution of a continuous variable

The normal curve serves as a good approximation for the histograms obtained from many distributions of continuous variables.

Example 1
A brand of electric light bulbs has an average life of 1 year and a standard deviation of 3 months. What is the probability that a bulb chosen at random will have a life of at least 18 months?

Solution

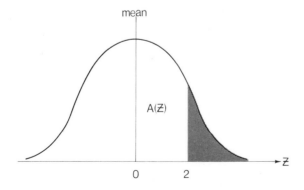

Fig. 11.19

124

$$\mu = 12 \text{ months}$$
$$\sigma = 3 \text{ months}$$
$$x = 18 \text{ months}$$

$$Z = \frac{x - \mu}{\sigma}$$

$$= \frac{18 - 12}{3}$$

$$= \frac{6}{3}$$

$$= 2$$

$$\therefore A(Z) = 0.477$$

$$\text{Required area} = 0.5 - 0.477$$

$$= 0.023$$

i.e. the probability of the bulb having a life of at least 18 months (18 months or more) is 0.023.

Example 2
Sacks of grain packed by an automatic machine loader have an average weight of 114 kg. It is found that 10% of the bags are over 116 kg. Find the standard deviation.

Solution

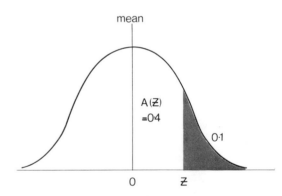

Fig. 11.20

In this type of example we must start with the area under the curve and work back to Z.
We know 10% of the bags are over 116 kg. This means the area to the right of Z is 10% or 0.1.

$$A(Z) = 0.5 - 0.1$$
$$= 0.4$$
$$\therefore Z = 1.28$$

Now we know, $\mu = 114\,\text{kg}$.
$$\sigma = ?$$
$$x = 116\,\text{kg}.$$
$$Z = 1.28$$
$$Z = \frac{x - \mu}{\sigma}$$
$$\frac{1.28}{1} = \frac{116 - 114}{\sigma}$$
$$= \frac{2}{\sigma}$$
$$1.28\sigma = 2$$
$$\sigma = \frac{2}{1.28}$$
$$= 1.56 \text{ (correct to 2nd decimal place)}$$

i.e. the standard deviation is 1.56

Example 3
Given a normal distribution of a continuous variable (x) with 2000 variates, the mean and standard deviation being 20 and 5 respectively, find the number of variates between x = 12 and x = 22.

Solution

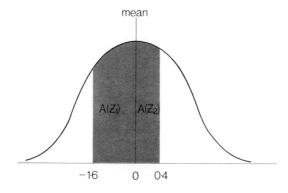

Fig. 11.21

$$N = 2000$$
$$\mu = 20$$
$$\sigma = 5$$
$$x_1 = 12$$
$$x_2 = 22$$
$$Z_1 = \frac{x_1 - \mu}{\sigma}$$
$$= \frac{12 - 20}{5}$$
$$= -1.6$$
$$Z_2 = \frac{x_2 - \mu}{\sigma}$$
$$= \frac{22 - 20}{5}$$
$$= 0.4$$
$$\therefore A(Z_1) = 0.445$$
$$\text{and } A(Z_2) = 0.155$$
$$\therefore \text{Required area} = 0.445 + 0.155$$
$$= 0.600$$

i.e. the probability of x lying between 12 and 22 is 0.6
∴ The expected number of variates between 12 and 22 = 2000 × 0.6
$$= 1200$$

EXERCISE D

1. A hundred standard squash balls are tested by dropping from a height of 250 cm and measuring the height of bounce. A ball is 'fast' if it rises above 80 cm. The average height of bounce was 75 cm and the standard deviation was 1.9 cm. What is the chance of getting a 'fast' standard ball?
2. Sacks of grain packed by an automatic loader have an average weight of 61 kg. It is found that 10% of the bags are below 59 kg. Find the standard deviation.
3. In the previous example, the machine is adjusted and the average weight per bag is now 60 kg. Assuming the standard deviation

is unaltered calculate the probability that a bag is now below 59 kg.

4. Suppose your score in an examination in standard units is 1.2 and the scores are assumed to be normally distributed, what percentage of the students would be expected to score higher than you?

5. Two brands of torch batteries have the same average life of 55 hours but different standard deviations of $1\frac{1}{4}$ hours and $1\frac{3}{4}$ hours. In each case, what is the chance that a battery will burn at least 58 hours?

6. A physical education teacher gives grades relative to all his classes for athletic events. If experience has shown that the average height is 178 cm and the standard deviation is 10 cm in the high jump, and the teacher gives 20% A's, how high would a pupil need to jump to expect an A?

7. If a normal distribution of a continuous variable has mean 21.2 and standard deviation 3.10, find the probability that a variate selected at random will be larger than 30 or less than 15.

8. The weight of grapefruit from a large shipment averages 420 g with a standard deviation of 42 g. If these weights are normally distributed, what percentage of all these grapefruit would be expected to weigh between 420 and 476 g?

9. If the average life of a certain make of storage battery is 30 months with a standard deviation of 5 months, what percentage of these batteries can be expected to last 24 to 36 months?

10. Records indicate that the average life of T.V. tubes is 3 years 6 months and the standard deviation is 18 months. Tubes lasting less than a year are replaced free. For every 200 sets (one tube to a set) sold, how many tubes can be expected to have to be replaced free?

11. The average yearly rainfall in a certain town is 75 cm with a standard deviation of 25 cm. Assuming this yearly rainfall follows a normal distribution, in how many years out of a period of 100 years could the residents of the town expect less than 37.5 cm of rain?

12. If the heights of 1000 male college students closely follow a normal distribution with a mean of 172.5 cm and a standard deviation of 6.25 cm.
   a) how many of these students would you expect to be at least 180 cm tall?
   b) what range of heights would include the middle 50% of the men in this group?

13. In grading a certain type of plum, 20% are called small, 55% medium, 15% large and 10% very large. If the weights follow a normal distribution, and the average weight is 138 g with a standard deviation of 33.5 g, what are the lower and upper bounds for the weight of medium plums?

14. In a normal distribution with a standard deviation of 2, the probability that a variate selected at random exceeds 26 is 0.05. Find the mean of the distribution and the variate above which lie 80% of all variates of the distribution.

15. In an examination, the average mark was 60 and the standard deviation 10. The teacher gave 30 students with marks between 51 and 69 a grade of C. If the marks are assumed to follow a normal distribution, how many students sat the examination?

16. A normal distribution has a mean of 30 and a standard deviation of 7. If it is known that 346 variates exceed 32.5, what is the total number of variates in the distribution?

17. A teacher who gives 10% A's, 20% B's, 40% C's, 20% D's and 10% E's, sets an examination in which the average mark is 54. If the borderline between the C's and the

126

D's is 44, what is the standard deviation in that examination?

18. In a normal distribution with mean 100 and standard deviation 44, there are 150 variates greater than 180. How many variates should we expect between 120 and 180?

19. In a population of adult men 20% are 175 cm or over in height and 10% are less than 155 cm. Find the mean and standard deviation of the population.

20. In an examination 10% of the class receive a grade of A, 20% B, 55% C and 15% D. The C grade ranges from 55 to 70. Assuming a normal distribution, what are the mean and standard deviation of the marks?

The binomial distribution as an approximation to the normal distribution

If a frequency distribution is obtained by using the terms of the binomial expansion, it is called a binomial distribution, e.g. the expected frequencies of the occurrence of heads in tossing N coins follow a binomial distribution.

The histograms representing the expected frequencies of heads in 64 tosses of 1, 2, 3, 4, 5 and 6 coins are shown below.

If the number of the sample, N, is large the calculation of frequencies and probabilities by means of the binomial theorem becomes tedious. Since many practical problems involve samples of large size, it is important to find a more rapid method of calculating probabilities. Such a method is furnished by the normal distribution which is the most important continuous probability distribution.

From the figures in Figure 11.22 it is apparent as N increases, the tops of the rectangles of the histograms approach a bell-shaped curve. This limiting frequency curve obtained as N becomes larger and larger is called the normal frequency curve (or normal probability curve). The curve approaches the

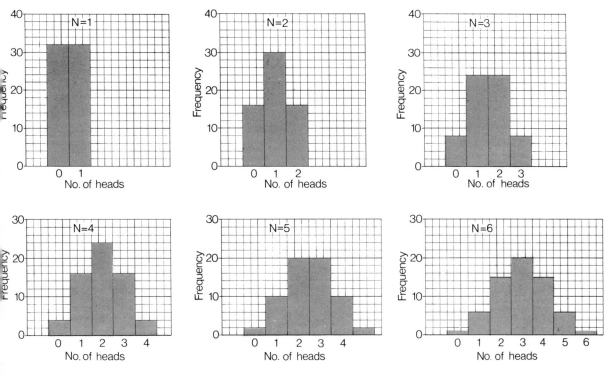

Fig. 11.22

127

horizontal axis but never touches it. The total area under the normal probability curve representing the probability that x will fall anywhere is equal to 1.

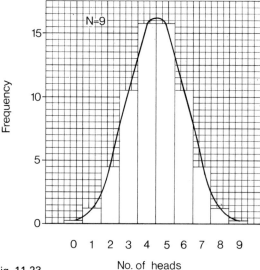

Fig. 11.23

No. of heads

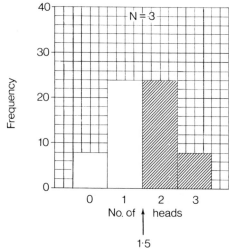

Fig. 11.24

Hence, by using the tables for finding the areas under the standard normal curve, we can find the probability of a certain event occurring.

To calculate the standard score (or Z score), we must know the mean and standard deviation of the distribution. The proof of the values of these measures is beyond our scope and here we shall only quote the results. If p is the probability of the success of an event in a *single trial* and q is the probability of its failure, then the binomial distribution giving the expected frequencies for 0, 1, 2 ------ N successes in N trials has

$$\text{mean } \mu = Np$$

and standard deviation $\sigma = \sqrt{Npq}$

Note: The binomial distribution deals with *discrete* variables only e.g. the number of times a head is obtained when tossing a coin, the number of times a 6 is obtained with a die, the number of defective T.V. tubes or light bulbs etc. This means that in working examples, if we wish to include a certain score we have to adjust the x value used.

128

Let us look at the third histogram shown in Figure 11.22 of the expected number of heads obtained in 64 tosses of 3 coins. If we wish to know how many times we obtained 2 heads or more, then to include 2 heads we must use the lower boundary of the '2' column which is 1.5. If we want to know how many times we obtained 1 or less heads we must use the upper boundary of the '1' col. which is 1.5.

Example 1
Using the normal curve approximation, find the probability that 3–6 heads are obtained in a toss of 9 coins (i.e. the probability that x lies between and *includes* 3 and 6).

Solution

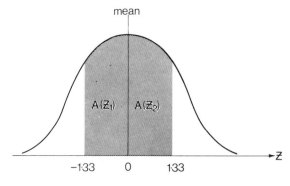

Fig. 11.25

$p = \frac{1}{2}$ = heads

$q = \frac{1}{2}$ = tails

$N = 9$

$\mu = Np$

$\quad = 9 \times \frac{1}{2}$

$\quad = 4.5$

$\sigma = \sqrt{Npq}$

$\quad = \sqrt{9 \times \frac{1}{2} \times \frac{1}{2}}$

$\quad = 1.5$

$x_1 = 2.5$ (to include 3 heads)

$x_2 = 6.5$ (to include 6 heads)

$Z_1 = \dfrac{x_1 - \mu}{\sigma}$

$\quad = \dfrac{2.5 - 4.5}{1.5}$

$\quad = \dfrac{-2}{1.5}$

$\quad = -1.33$

$A(Z_1) = 0.408$

$Z_2 = \dfrac{x_2 - \mu}{\sigma}$

$\quad = \dfrac{6.5 - 4.5}{1.5}$

$\quad = \dfrac{2}{1.5}$

$\quad = 1.33$

$A(Z_2) = 0.408$

Required area $= 0.408 + 0.408$

$\qquad\qquad = 0.816$

i.e. the probability of obtaining 3—6 heads is 0.816.

Example 2
If 12 dice are thrown, what is the probability, using the normal curve approximation, that 6 or more dice will show a 5?

Solution

$p = \frac{1}{6}$ = a 5

$q = \frac{5}{6}$ = any other number.

$N = 12$

$\mu = Np$

$\quad = 12 \times \frac{1}{6}$

$\quad = 2$

$\sigma = \sqrt{Npq}$

$\quad = \sqrt{12 \times \frac{1}{6} \times \frac{5}{6}}$

$\quad = 1.29$

$x = 5.5$ (to include 6 fives)

$Z_1 = \dfrac{x - \mu}{\sigma}$

$\quad = \dfrac{5.5 - 2}{1.29}$

$\quad = \dfrac{3.5}{1.29}$

$\quad = 2.71$

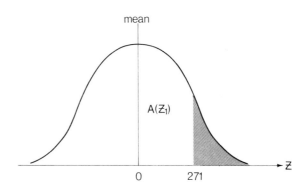

Fig. 11.26

$A(Z_1) = 0.497$

Required area $= 0.5 - 0.497$

$\qquad\qquad = 0.003$

i.e. the probability of obtaining 6 or more fives is 0.003.

129

Example 3

A manufacturer producing tea cups finds that 10% of them are defective. What is the probability that in a sample of 50 cups, no more than 4 are defective?

Solution

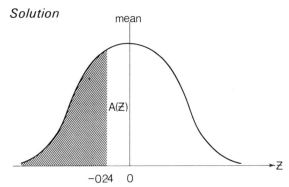

Fig. 11.27

$$p = \frac{1}{10} = \text{defective}$$

$$q = \frac{9}{10} = \text{not defective}$$

$$N = 50$$

$$\mu = Np$$

$$= 50 \times \frac{1}{10}$$

$$= 5$$

$$\sigma = \sqrt{Npq}$$

$$= \sqrt{50 \times \frac{1}{10} \times \frac{9}{10}}$$

$$= 2.12$$

$$x = 4.5 \text{ (to include 4)}$$

$$Z = \frac{x - \mu}{\sigma}$$

$$= \frac{4.5 - 5}{2.12}$$

$$= \frac{-0.5}{2.12}$$

$$= -0.24$$

$$A(Z) = 0.095$$

Required area $= 0.5 - 0.095$

$$= 0.405$$

i.e. the probability of obtaining no more than 4 (4 or less) defectives is 0.405.

EXERCISE E

1. A coin is tossed 100 times. Use the normal curve approximation to find the probability of obtaining
   a) 60 or more heads
   b) exactly 50 heads

2. Use the normal curve approximation to find the probability of obtaining exactly 16 sixes in 96 tosses of a die.

3. For a binomial frequency for which $p = \frac{1}{4}$ find the probability of obtaining 25 or more successes in 80 trials.

4. If 40% of students have defective eyesight, what is the probability that at least half of the members of a class of 50 students will have defective eyesight? Use the normal curve approximation.

5. If 10% of T.V. tubes burn out before their guarantee has expired,
   a) what is the probability that a shopkeeper who has sold 200 such tubes will have to replace at least 30 of them?
   b) what is the probability that he will replace at least 10 and not more than 30 tubes.
Use the normal curve approximation.

6. If 25% of the drivers in a certain town have at least one accident in a year's driving, what is the probability that 30% or more of 500 customers of an insurance company will have an accident during the next year? Use the normal curve approximation.

7. Assume that half of the people in a certain community are regular viewers of T.V. Of 100 investigators, each interviewing 10

130

individuals regarding their viewing habits, how many would you expect to report that 3 people or fewer were regular T.V. viewers? (Use normal distribution.)

8. A manufacturer of light bulbs finds that on average 5% are defective. What is the probability that out of 1000 such bulbs selected at random 30 or more are defective?

9. A pair of dice is rolled 25 times. What is the probability that a 7 will show 10 or more times?

10. Within a certain age group of children it is known that 20% require dental attention. A sample of 80 such children are to be studied.
   a) Calculate the mean and standard deviation of the number of children in such a sample who require dental attention.
   b) Assuming that the distribution is approximately normal calculate the probability that
      1) 20 or more children require dental attention.
      2) 15 to 20 inclusive of the children require dental attention.

11. An examination consists of 40 multiple choice questions, each with 4 possible responses, of which 1 is correct. If a candidate is completely unprepared for the examination and decides to choose his answers at random, find the mean and standard deviation of the number of questions answered correctly.
   Using the normal distribution as an approximation to the distribution, calculate the probability that he answers correctly 6 or fewer questions. In the examination, 18 or more questions answered correctly gained a pass. On the above basis find how many, out of a group of 500 students who chose to select their answers at random, would be expected to gain a pass.

12. A factory making ball point pens is experimenting with a new method of production. 5% of the pens are rejected as below standard. A sample of 50 pens is examined.
   a) Calculate the mean and standard deviation of the number of pens rejected in such a sample.
   b) Given that the distribution of the number of rejected pens is approximately normal, calculate the probability that
      1) 3 or fewer pens are rejected,
      2) from 2 to 5 pens are rejected.

13. According to a certain insurance company, the probability of a man aged 45 dying within a year is 0.05. If the company has 20 000 policies in force on men of this age estimate the probability that the company will have to pay out more than 900 death claims on this group of men within a year.

14. A manufacturer of radio valves knows that on average 2% of his products are defective. Using the normal curve approximation, what is the probability that a sample of 100 valves will contain exactly 5 defective valves?

15. How many times would 2 dice have to be rolled so that there is a 50% chance of getting at least 3 fives?

12 SAMPLE MEASURES

In Chapter 3 we looked at the problem of obtaining a satisfactory sample. We would now consider some measures pertaining to samples and their relationship with the measures of the parent population from which the sample was taken.

Notation used for samples and populations

The **population** is the general term given to the aggregate of whatever we are considering. It only means a human population when people are being considered. A **sample** is a part of the population.

The following notation is used:

	Sample	Population
Number of variates	N	n
Mean	$\overline{X}$	μ
Standard deviation	s	σ

A numerical characteristic of a population such as its mean or standard deviation is called a **parameter.**

A quantity calculated from a sample such as its mean and standard deviation is called a **statistic.**

Since parameters of a given population are based upon *all* its variates, they are fixed for that population. On the other hand, since statistics are based upon only a *part* of the population, they usually vary from sample to sample.

Sampling distributions

The results of an arithmetic test for speed and accuracy given to all first year pupils (179) in a secondary school are shown below.

Mean numbers of sums correct = 4.14.

We wished to use this distribution for sampling purposes. Using random numbers the scores for 40 of the pupils were selected and the mean was calculated as shown below. These results were then returned to the distribution and the procedure repeated until 50 samples in all had been taken (all samples of size 40). In each case the mean of the sample was calculated.

Sample 1

No. of Sums correct (X)	Freq. (f)	fX
0	0	0
1	2	2
2	6	12
3	8	24
4	6	24
5	7	35
6	7	42
7	1	7
8	2	16
9	1	9
10	0	0
	$\sum f = 40$	$\sum fX = 171$

$$\overline{X} = \frac{\sum fX}{f}$$

$$= \frac{171}{40}$$

$$= 4.275$$

approx. 4.28

Mean of Sample 1 = 4.28

Number of sums correct	0	1	2	3	4	5	6	7	8	9	10
Frequency	4	10	20	33	39	29	25	13	4	2	0

The means of the samples (or sample means) were formed into a frequency distribution—a **sampling distribution** as shown below.

SAMPLING DISTRIBUTION

Sample Means	Frequency
3.50–3.59	1
3.60–3.69	1
3.70–3.79	3
3.80–3.89	3
3.90–3.99	5
4.00–4.09	5
4.10–4.19	6
4.20–4.29	8
4.30–4.39	10
4.40–4.49	2
4.50–4.59	4
4.60–4.69	2

The mean of this sampling distribution was calculated and found to be 4.17 i.e. the mean of the sample means was 4.17.

Distribution of the sample means

A variable X has a normal distribution with mean μ and standard deviation σ.

If observations are collected, not individually but as random samples of size N, then the means of the samples can be formed into a frequency distribution which will be approximately normal. This distribution is called a sampling distribution, the variable this time being $\bar{X}$.

Fig. 12.2

Fig. 12.1 Sampling Distribution

This histogram shows that the sampling distribution is reasonably near to a normal distribution considering that the number of samples is only 50.

The mean of the sampling distribution (or mean of the sample means) is 4.17, very close to the mean of the population 4.14.

Try an experiment like this yourselves with some data of your own. You should use a distribution which is approximately normal with a fairly large total frequency.

The mean of the sampling distribution is called $\mu_{\bar{x}}$ to distinguish it from the population mean μ. Similarly the standard deviation of the sampling distribution is called $\sigma_{\bar{x}}$.

There is a definite correlation between the mean and standard deviation of the sampling distribution, and the mean and standard deviation of the population.

For practical purposes we can assume that the mean of the sampling distribution (i.e.

133

the mean of the sample means) $\mu_{\bar{x}}$ is approximately the same as the population mean.

i.e. $\mu_{\bar{x}} = \mu$

Also, $\sigma_{\bar{x}}$ the standard deviation of the sampling distribution is equal to the population standard deviation, σ, divided by the square root of N, the number in the samples.

i.e. $\sigma_{\bar{x}} = \dfrac{\sigma}{\sqrt{N}}$

This standard deviation, $\sigma_{\bar{x}}$, of the sample mean $\bar{X}$ from the population mean μ is known as the **standard error** of the sample mean from the population mean.

i.e. The Standard Error $= \dfrac{\sigma}{\sqrt{N}}$

Obviously as N, the size of the sample, increases, the standard error of the sample mean from the population mean i.e. the difference between the sample mean and the population mean, becomes smaller and the standard deviation, s, of a single *large* sample approximates to, σ the standard deviation of the population i.e. s $= \sigma$.

We standardise a score X of a normal distribution with mean μ and standard deviation σ by using this formula.

Standard score, $Z = \dfrac{X - \mu}{\sigma}$

In the same way we may standardise a score $\bar{X}$ of the sampling distribution with mean $\mu_{\bar{x}}$ and standard deviation $\sigma_{\bar{x}}$ by using the same formula in this way:

Standard score, $Z = \dfrac{X - \mu_{\bar{x}}}{\sigma_{\bar{x}}}$

Example 1
The mean height of 1000 university students is 171.25 cm and the standard deviation is 6.25 cm. Find the probability that in a sample of 100 students, the mean will be greater than 172.50 cm.

Solution

Population: $\mu = 171.25$ cm

$\sigma = 6.25$ cm

Sampling: $\mu_{\bar{x}} = \mu = 171.25$ cm

$\sigma_{\bar{x}} = \dfrac{\sigma}{\sqrt{N}}$

$= \dfrac{6.25}{\sqrt{100}}$

$= \dfrac{6.25}{10}$

$= 0.625$ cm

We must find the area under the normal curve to the right of $\bar{X} = 172.50$ (the mean of the sample of 100 students).

$$Z = \dfrac{\bar{X} - \mu_{\bar{x}}}{\sigma_{\bar{x}}}$$

(We are standardising the sample mean $\bar{X} = 69$ in the sampling distribution)

$= \dfrac{172.50 - 171.25}{0.625}$

$= \dfrac{1.25}{0.625}$

$= 2$

Fig. 12.3

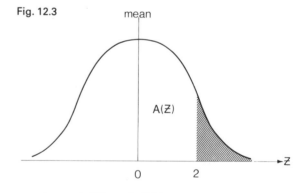

$\therefore A(Z) = 0.477$

$\therefore$ Reqd. area $= 0.5 - 0.477$

$= 0.023$

i.e. The probability of the sample mean being greater than 172.50 cm $= 0.023$

134

Example 2
Assume that the heights of 3000 men are normally distributed with mean 172.5 cm and standard deviation 7.5 cm. If 50 samples of 25 men each (assuming replacement) are obtained, what would be the expected mean and standard deviation of the resulting sampling distribution? In how many samples would you expect to find the mean between 168.75 and 173.75 cm?

Solution

$$\text{Population:} \quad \mu = 172.5 \text{ cm}$$

$$\sigma = 7.5 \text{ cm}$$

$$\text{Sampling:} \quad \mu_{\bar{x}} = \mu = 172.5 \text{ cm}$$

$$\sigma_{\bar{x}} = \frac{\sigma}{\sqrt{N}}$$

$$= \frac{7.5}{\sqrt{25}}$$

$$= \frac{7.5}{5}$$

$$= 1.5 \text{ cm}$$

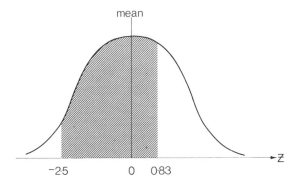

Fig. 12.4

We must find the area under the normal curve

$$\text{from } \bar{X}_1 = 168.75 \text{ to } \bar{X}_2 = 173.75$$

$$Z_1 = \frac{\bar{X}_1 - \mu_{\bar{x}}}{\sigma_{\bar{x}}}$$

$$= \frac{168.75 - 172.50}{1.5}$$

$$= \frac{-3.75}{1.5}$$

$$= -2.5$$

$$Z_2 = \frac{\bar{X}_2 - \mu_{\bar{x}}}{\sigma_{\bar{x}}}$$

$$= \frac{173.75 - 172.50}{1.5}$$

$$= \frac{1.25}{1.5}$$

$$= 0.83$$

$$A(Z_1) = 0.494$$

$$A(Z_2) = 0.297$$

$$\text{Reqd. area} = 0.494 + 0.297$$

$$= 0.791$$

∴ Probability of the sample mean being between 168.75 and 173.75 cm = 0.791
Expected number of samples

$$= \text{prob.} \times \text{no. of samples}$$

$$= 0.791 \times 50$$

$$= 39.55$$
$$= \text{approx. } 40$$

EXERCISE A

1. In a normal distribution with mean 73.5 and standard deviation 2.80, find the probability that in a sample of 100 variates, the mean will be
   a) more than 74
   b) less than 72.6

2. In a normal distribution with mean 53.6 and standard deviation 3.50, find the probability that in a sample of 36 variates, the mean will be
   a) less than 52.8
   b) between 53.2 and 54.

3. In a savings bank, the average account is £160.85 with a standard deviation of £18. What is the probability that a group of 81 accounts taken at random, will show an average deposit of £165 or more?

4. The heights of a certain group of adults are normally distributed with a mean of 170.75 cm and a standard deviation of 6.25 cm. If 25 people are chosen at random from the group, what is the probability that their mean height will be 172.25 cm or more?

5. If all possible samples of size 36 are drawn from a normally distributed population with mean 30 and standard deviation 3, within what range will the middle 50% of the sample means lie?

6. The weekly wages of a certain industry are normally distributed with a mean of £20. If 10% of the mean wages of samples of 25 workers fall below £19.50, what is the standard deviation of weekly wages in this industry?

7. (a) Assuming that the heights of men are normally distributed with a standard deviation of 5 cm, how large a sample should be taken in order to be fairly sure (with a probability of 0.95) that the sample mean does not differ from the population mean by more than 1.0, in absolute value?

(b) How large a sample would need to be taken to be 99% sure?

8. A normal population has a standard deviation of 3. How large a sample should be taken in order to be 99% sure that the sample mean differs from the true mean by less than 0.5?

Confidence limits

Suppose we have a sample of N variates with mean $\bar{x}$ and standard deviation s selected from a normal population. Let us consider the problem of estimating the mean of the population from this single sample. It is obviously impossible to determine the population mean precisely, since this sample could have been taken from any population with roughly the same size of mean. What we can do, is establish **limits** within which the mean of the population will fall with a specified probability or **confidence.**

136

To illustrate this, suppose a random sample of 100 variates is taken from a normal population. The mean of the sample is found to be 30 and the standard deviation is 5. From this one sample we will try to estimate the mean of the population with a probability of 95% (or 95% confidence).

With a probability of 0.95 the population mean will fall in the shaded area of the diagram.

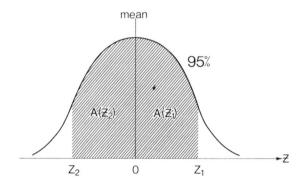

Fig. 12.5

i.e. $A(Z_1) = 0.475$

$$A(Z_2) = 0.475$$

$$\mu_{\bar{x}} = \mu$$

$$\sigma_{\bar{x}} = \frac{\sigma}{\sqrt{N}}$$

We do not know either $\sigma_{\bar{x}}$ or σ. However, the standard error, $\frac{\sigma}{\sqrt{N}}$, of the sample mean from the population mean i.e. the difference between the sample mean and the population mean becomes smaller as N becomes larger, so that for a **large sample,** we may assume that the standard deviation of the sample, s, approximates to σ.

Thus $\sigma_{\bar{x}} = \frac{\sigma}{\sqrt{N}}$

$$= \frac{s}{\sqrt{N}} = \frac{5}{\sqrt{100}} = \frac{5}{10} = 0.5$$

From the tables we find that

$$Z = +1.96$$
$$Z = -1.96$$

i.e. we must say

$$Z = \pm 1.96$$

$$Z = \frac{\bar{X} - \mu_{\bar{x}}}{\sigma_{\bar{x}}}$$

$$\pm 1.96 = \frac{30 - \mu}{0.5}$$

$$\therefore \ \pm 1.96 \times 0.5 = 30 - \mu$$

$$\therefore \ \mu = 30 - (\pm 1.96 \times 0.5)$$
$$= 30 \pm 1.96 \times 0.5$$
$$= 30 \pm 0.980$$

With 95 % confidence we can say that the population mean lies within the limits 30 ± 0.980.

(The answer is very often left in this form)

We could also have given the answer like this:

The 95 % confidence limits for the population mean are 30 ± 0.980.

(These limits define the interval which would contain the population mean with a probability of 95 %).

The interval between the confidence limits is called the confidence interval.

Fig. 12.6

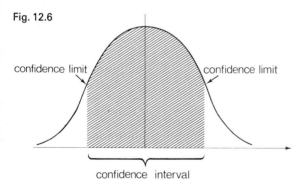

With 95 % confidence limits, we find Z_1 and Z_2 by using the tables to find the Z score corresponding to these areas

$$A(Z_1) = \tfrac{1}{2} \text{ of } 0.95 = 0.475$$
$$A(Z_2) = \tfrac{1}{2} \text{ of } 0.95 = 0.475$$

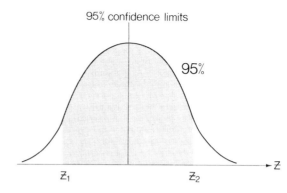

Fig. 12.7 (a) 95% Confidence Limits

With 99 % confidence limits, we find Z_1 and Z_2 by using these areas.

$$A(Z_1) = \tfrac{1}{2} \text{ of } 0.99 = 0.495$$
$$A(Z_2) = \tfrac{1}{2} \text{ of } 0.99 = 0.495$$

Similarly we can find Z for any specified confidence.

Fig. 12.7 (b) 99% Confidence Limits

Note: We can only use the above method for estimating a population mean when the sample is large, where it is permissible to assume that the standard deviation of the sample can be taken as an approximation of the standard deviation of the population.

Example

A sample of 80 variates has a mean of 60 with a standard deviation of 5.5. Find the 98 % confidence limits for the mean of the population.

137

Solution

Here, $\bar{X} = 60$

$s = 5.5$

$N = 80$

And, $\mu_{\bar{x}} = \mu$

$$\sigma_{\bar{x}} = \frac{\sigma}{\sqrt{N}}$$

$$= \frac{s}{\sqrt{N}} \quad \text{(Since the sample is large} \atop \text{σ can be taken as s)}$$

$$= \frac{5.5}{\sqrt{80}} = \frac{5.5}{8.94} = 0.62$$

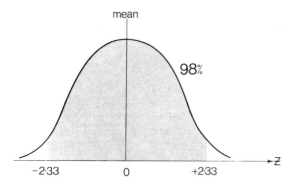

Fig. 12.8

With a probability of 0.98

$$A(Z_1) = \tfrac{1}{2} \text{ of } 0.98 = 0.49$$

and $A(Z_2) = 0.49$

$$\therefore Z_1 = +2.33$$

$$\therefore Z_2 = -2.33$$

i.e. $Z = \pm 2.33$

$$Z = \frac{\bar{X} - \mu_{\bar{x}}}{\sigma_{\bar{x}}}$$

$$\therefore \pm 2.33 = \frac{60 - \mu}{0.62}$$

$$\therefore \pm 2.33 \times 0.62 = 60 - \mu$$

$$\therefore \mu = 60 \pm 2.33 \times 0.62$$

$$= 60 \pm 1.4446$$

$$= 60 \pm 1.44$$

i.e. The 98% confidence limits for the population mean are

$$60 \pm 1.44$$

EXERCISE B

1. In a sample of 60 variates, the mean is 30 and the standard deviation 3.8. Find the limits within which the population mean is expected to lie, with a probability of 95%.

2. A sample of 100 variates has a mean of 65 and a standard deviation of 2.5. Find the 99% confidence limits for the mean of the population.

3. Find the 98% confidence limits for the mean of a population, if a sample of 90 variates taken from the population has a mean of 12.6 and a standard deviation of 3.

4. Suppose the standard deviation of heights in males is 6.25 cm. Two hundred male students in a large university are measured and their average height is found to be 171.05 cm. Within what range of heights would you expect to find the mean height of the men in this university with a confidence of 95%?

5. In the same university, the heights of 100 women students were measured and the mean height was found to be 161.25 cm with a standard deviation of 5.5 cm. With 95% confidence, find the limits within which the mean height of the women in that university lie.

6. A sample of 50 light bulbs was taken from a very large batch produced in a factory. The mean life of the bulbs was found to be 2.5 years with a standard deviation of 6 months. Estimate the mean life of the batch of bulbs with a probability of 99%.

ANSWERS

Chapter 1
Exercise C   Page 9

2. *a*) 5800   *b*) 3000   *c*) 48%   *d*) $\frac{20}{23}$ *e*) $\frac{3}{35}$

5. *a*) 1st Friday   *b*) 1st Monday and 1st Tuesday   *c*) 1st Friday and 2nd Monday

Exercise F   Page 16

1. *a*) Mary   *b*) Mary   *c*) Anne, Joan, Mary

3. £7 200 000; Germany £1 800 000; U.S.A. £1 000 000; Israel £400 000; Denmark £1 600 000.

5. The sales scale does not start at 0 and any rises appear greater. The time scale is not regular.

6. The lengths of the "blocks of butter" are drawn proportionally to the weights but the apparent volume of the blocks is in a quite different proportion.

7. The profits scale does not start at 0 and so rises appear greater. The last part of the line 1980–90 shows imaginary future profits with no apparent foundation for the optimistic hope.

8. The diameters of the "suns" are in proportion to the hours of sunshine, but the area shown by the circles is not in the same ratio.

9. The time scale is not regular. The points are not plotted to scale with regard to the profits figures.

Chapter 2
Exercise A   Page 19

1. *a*) 5.63   *b*) 10.1   *c*) 10   *d*) 0.0068   *e*) 6   *f*) 120
2. *a*) 11.7   *b*) 9.8   *c*) 69.0   *d*) 0.1   *e*) 7.0   *f*) 15.8
   g) 15.8 or 15.9   *h*) 15.9
3. *a*) 600,000   *b*) 580,000   *c*) 583,000   *d*) 582,700   *e*) 582,730

Exercise B   Page 20

	a)	*b*)	*c*)
1.	1 cm	0.5 cm	7.5 cm, 6.5 cm
2.	1 m	0.5 m	186.5 m, 185.5 m
3.	1 h	0.5 h	17.5 h, 16.5 h
4.	0.1 kg	0.05 kg	9.25 kg, 9.15 kg
5.	0.1 g	0.05 g	12.85 g, 12.75 g
6.	0.001 litres	0.0005 l	2.6835 l, 2.6825 l
7.	0.01 mile	0.005 mile	1.215 mile, 1.205 mile
8.	0.1 s	0.05 s	16.55 s, 16.45 s

Exercise C   Page 21

	a)	b)		a)	b)
1.	0.5 m	$\dfrac{1}{300}$	6.	$\dfrac{1}{70}$	1.4%
2.	0.5 kg	$\dfrac{1}{70}$	7.	$\dfrac{1}{34}$	2.9%
3.	0.05 l	$\dfrac{1}{30}$	8.	$\dfrac{1}{500}$	0.20%
4.	0.05 s	$\dfrac{1}{46}$	9.	$\dfrac{1}{2500}$	0.040%
5.	$\dfrac{1}{14}$	7.1%	10.	$\dfrac{1}{30\,000}$	0.0033%

Exercise D   Page 21

1. *a*) 14.80 cm,   14.60 cm   *b*) 33.0 g,   31.0 g
   c) 41.0 m,   39.0 m   *d*) 40.0 litres,   38.0 litres
2. *a*)   1.0 m   *b*) 1.0 kg   *c*) 0.10 g   *d*) 0.010 cm
3. *a*) 15.0 cm,   13.0 cm   *b*) 6.0 kg,   4.0 kg
   c) 7.40 m,   7.20 m   *d*) 3.70 litres,   3.50 litres
4. *a*)   1.0 m   *b*) 1.0 ml   *c*) 0.55 cm   *d*) 0.55 g

Exercise E   Page 22

1.	503.75	2.	19.65	3.	34.77	4.	165.1425
5.	10.47	6.	9.30	7.	14.75	8.	0.43
9.	0.65	10.	8.11				

Chapter 3
Exercise A   Page 25

1. *b*)

Year	1	2	3	4	5	6
No. of pupils	24	24	26	14	8	4

2.

Age (years)	0–4	5–15	16–45	46–65	65+
No. of people	3	9	10	15	3

3.

Class	1	2	3	4	5	6	7
No. of pupils	6	6	7	7	7	9	8

4. Numbers for sample
295, 297, 281, 264, 167, 276, 272, 35, 132, 107

5. Numbers for sample
29, 52, 79, 70, 2

6. *a*) Every 20th employee.
   b) Numbers for sample
356, 360, 87, 2, 153, 89, 112, 56, 23, 167, 295, 297, 254, 314, 379, 276, 252, 90, 430, 271, 332, 159, 65, 192, 412

7. *a*) Every 8th employee.
   b) Numbers for sample
35, 63, 60, 8, 70, 2, 59, 11, 53, 56

Chapter 4
Exercise A   Page 27

1. Quantitative — continuous
2. Quantitative — discrete
3. Qualitative
4. Quantitative — continuous
5. Qualitative
6. Quantitative — discrete
7. Quantitative — continuous
8. Quantitative — continuous
9. Quantitative — discrete
10. Quantitative — discrete
11. Quantitative — continuous
12. Qualitative
13. Quantitative — continuous.

Exercise B Page 29

1.

Heights	142	144	146	148	150	152	154	156	158	160	162	164
Frequency	1	1	0	3	5	8	10	4	6	1	0	1

a) Continuous   b) ·22 cm   c) 154 cm   d) 0.25 or 25%
e) 5   f) 5%   g) 67.5%

3.

No. of children	1	2	3	4	5	6	7	8	9	10
Frequency	2	4	5	2	4	7	1	1	1	2

a) Discrete   b) 6 children   c) Families with no. children
d) 5   e) 37.9%

5. a)

Goals	0	1	2	3	4	5
Frequency	7	11	10	4	3	1

b) Discrete   c) 5 goals   d) 1 goal   e) 0.306 or 30.6%
f) 22.2%

Exercise C Page 30

1. a) 22   b) 13   c) 16   d) 0.18   e) 6   f) 6
2. a) The half sizes do not have as large frequencies as might be expected. We discovered that this trend was markedly obvious all through the school for all age groups of boys and girls. To try to explain this, some pupils visited shoe shops in the area and interviewed the managers. The two main explanations for this trend were that mothers were inclined to buy the next whole size of shoe to allow for growth, rather than take the half size; and that the cheaper make of shoes are often only made in whole sizes.

b) 68   c) 5   d) 0.29 or 29%   e) 0.103 or 10.3%

4.

No. of Words	1	2	3	4	5	6	7	8	9	10	11	12	13
Frequency	1	0	2	0	2	4	3	11	7	3	3	1	1

a) 38   b) 8   c) 12   d) 0.18   e) 8
f) 3   g) 18

6. a) 32   b) 4   c) 8   d) 21.9%

Exercise D Page 32

1. a) 30   b) 8   c) 76   4. a) 3, 6

Exercise E Page 34

1.

Mark	1–5	6–10	11–15	16–20	21–25	26–30	31–35	36–40	41–45	46–50
Frequency	2	6	8	10	6	3	2	1	1	1

a) 23   b) 25.5   c) 5.5   d) 4th class   e) 0.2
f) 20%   g) 60%

3.

Rate	21–22	23–24	25–26	27–28	29–30	31–32	33–34	35–36	37–38
Frequency	2	7	10	8	9	5	1	1	1

a) 21.5   b) 26.5   c) 28.5   d) 3rd class   e) 0.2

4.

No. of rainy days	140–149	150–159	160–169	170–179	180–189	190–199	200–209
Frequency	1	4	4	7	11	10	9

No. of rain days	210–219	220–229	230–239	240–249	250–259	260–269
Frequency	2	5	4	3	2	3

a) 174.5   b) 189.5   c) 239.5   d) 5th class   e) 37   f) 56.9%

Exercise F   Page 36

1. *a)* 65   *b)* 50.5   *c)* 125.5   *d)* 88   *e)* 2nd class
   f) 0.4   *g)* 20%   *h)* 7.7%
2. *a)* 80   *b)* 49.5   *c)* 44.5   *d)* 45–49 kg   *e)* 55–59 kg
   f) 52   *g)* 4th class   *h)* 0.325

Exercise H   Page 41

1.

Heights (cm)	Cum. Freq.
146	1
146–148	1
−150	2
−152	5
−154	8
−156	12
−158	15
−160	20
−162	26
−164	27
−166	29
−168	30

a) 1
b) 60%
c) 3

2. *a)*

Time	Cum. Freq.
0– 20	0
− 40	2
− 60	6
− 80	13
−100	24
−120	52
−140	72
−160	88
−180	98
−200	104
−220	107

b) 6
c) 51.4%

3.

Weight (kg)	Cum. Freq.
35–39	1
35–44	3
35–49	6
35–54	14
35–59	22
35–64	26
35–69	27
35–74	27
35–79	28
35–84	29

a) 29
b) 6
c) 7

4. *a)*

Score	Freq.
65	1
66	2
67	2
68	8
69	7
70	7
71	5
72	2
73	2
74	2
75	2

b)

Score	Cum. Freq.
65	1
65–66	3
65–67	5
65–68	13
65–69	20
65–70	27
65–71	32
65–72	34
65–73	36
65–74	38
65–75	40

c) 20
d) 8

Exercise J   Page 42

1. *a)* Qualitative   *b)* Quantitative — discrete   *c)* Qualitative
   d) Quantitative — continuous

2. *a)*

Length (cm)	Frequency
0–19	0
−20	1
−21	4
−22	6
−23	10
−24	19
−25	27
−26	33
−27	38
−28	39
−29	39
−30	40

b)

Length (cm)	20	21	22	23	24	25	26	27	28	29	30
Frequency	1	3	2	4	9	8	6	5	1	0	1

c) 19   *d)* 7   *e)* 24 cm   *f)* 23.5 cm

3. *b)*

Interval (sec)	3	4	5	6	7	8	9	10	11	12	13
Frequency	2	1	3	1	2	2	2	4	2	2	1

   d) 10 sec   *e)* 13

4. *a)*

Weight (kg)	35–39	40–44	45–49	50–54	55–59	60–64	65–69	70–74
Frequency	0	1	2	6	4	3	3	1

   b) 50–54 kg   *c)* 45–49 kg   *d)* 44.5–49.5 kg

5.

Year	1	2	3	4	5
No. of pupils	150	162	156	138	114

6. *a)* 1600–1799 hr

   c)

Lifetime (hr)	Frequency
0–999	0
–1199	8
–1399	24
–1599	52
–1799	84
–1999	96
–2199	100
–2399	100

   d) 52

Chapter 5

Exercise A   Page **46**

1.

	Mean	Median	Mode
a)	approx 9.2	10	11
b)	23	22.5	21
c)	3	3	3

2. The mode, because it is the lowest.
3. *a)* 162 cm   *b)* 158 cm or 162 cm   *c)* 162.1 cm
6. The mode would be the easiest to find and the mean the hardest.
7. Mean—£6.80   Median £6.20

Exercise B   Page **47**

1. Approx. 1.25 goals
3. Approx. 3.9 children
5. Approx. 9.3 words per line

Exercise C   Page **48**

1. Approx. 56.4 marks
2. Approx. 39.4 years
3. Approx. 39.7 years
4. Approx. 77.4 hours
5. 23.8 sec
6. 479.8 g

Exercise D   Page **51**

1. 100   2. Approx. 146.4 cm   3. Approx. 116.08 g
4. 125.5 ships   5. 160.2 cm   6. 22.6 cm

Exercise E   Page **53**

1. *a)* 29.5   *b)* 23.5   2. *a)* 23.8 *b)* 55%   *c)* 45%
3. 158.1 cm   4. 122.5 secs   5. 55 kg
6. 69.7 strokes   7. 572 deaths.

Exercise F   **Page** 55

1. 6   2. 2.4   3. 3.1   4. 4.8   5. 4.4
6. 6.6   7. 3.7   8. 5.5   9. 5.2   10. 3.5

Exercise G   **Page** 55

1. *a*) 5, 5, 5   *b*) 79, 83, 90   *c*) 42, $35\frac{1}{2}$, 24
2. *a*) 6   *b*) 11.1   *c*) 6.4
3. *a*) 5.1, 5, 4   *b*) Mean
4. *a*) 2   *b*) 2.5   *c*) 5.7%   *d*) 41.5%
5. *b*) 49.9 min   *c*) 52 min   *d*) 24%
6. *a*) 21.9 kg   *b*) 21.7 kg   *c*) 18.5 kg   *d*) 77.5%
7. *a*) 3   *b*) 4.2   *c*) 4
8. *a*) Mode   *b*) All equal   *c*) Mean
9. *a*) 4.6   *b*) 54.5   *c*) 24.7 cm
10. 8.1 sec   11. *a*) 48 kg   *b*) 56.75 kg   12. 1544 hours

Chapter 6

Exercise A   **Page** 60

1. $13\frac{1}{3}$, 10, 9, 10, $11\frac{2}{3}$, 16, $16\frac{2}{3}$, $16\frac{2}{3}$.
2. 12.4, 10.0, 10.0, 14.0, 14.0, 14.6.
3. 23, $25\frac{1}{2}$, $25\frac{3}{4}$, 26, $28\frac{1}{4}$, $26\frac{1}{4}$, 28, $31\frac{1}{4}$, 32.
4. $193\frac{1}{2}$, $194\frac{1}{4}$, $189\frac{1}{2}$, $182\frac{3}{4}$, 178, $178\frac{1}{2}$, $183\frac{3}{4}$, $186\frac{1}{2}$, 181, 188, 194, $198\frac{3}{4}$, $211\frac{1}{4}$.
5. 204.3, 202.1, 199.6, 195.4, 194.8, 194.2, 196.1.
6. *a*) 1155, 1169, 1222, 1294, 1310, 1343, 1364, 1417, 1659.
   b) 1155, 1172, 1227, 1310, 1327, 1370, 1377, 1424, 1440.
   c) 1956, 2000, 2102, 2242, 2311, 2376, 2476, 2624, 2704.
   d) 1959, 2010, 2106, 2250, 2314, 2390, 2479, 2628, 2711.
7. £67, £68, £69, £68, £70, £71, £73, £74, £76, £75, £78, £85, £92, £92,
    £93, £96, £98, £100, £107, £107, £106, £104, £102, £96, £88.
8. 3.3%, 3.0%, 2.9%, 3.3%, 3.6%, 3.2%, 2.6%, 2.7%.
9. 14.6%, 14.6%, 14.4%, 14.5%, 14.3%, 14.0%, 13.5%, 13.0%, 12.6%.

Exercise B   **Page** 62

1. 8   2. 17   3. 50%   4. 7.5 gm per cm^3
5. 35.5   6. 63.6

Exercise C   **Page** 62

1. A 105   B 120   C 80   D 154   E 184
2. A £10.50 B £9.36   C £1.05   D £335   E £355
3. A 120   B 120   C 90   D 80   E 109
4. 147   5. 99

Exercise D   **Page** 64

1. 156   2. 119, 19%   3. 111   4. 114   5. 110

Chapter 7

Exercise A   **Page** 68

1.

	Range	*Semi-interquartile Range*
a)	6	1.5
b)	10	3
c)	10	3
d)	8	3

2. *a*) Median = 70.25 kg   *b*) Interquartile range = 2.8 kg   *c*) 6 kg
   d) 27 boys   *c*) 22 boys.
3. *a*) 46 marks; 64 marks   *b*) 9.2 marks; 12.7 marks
   c) 76 pupils   *d*) 45.7%   *e*) 65 marks.

4. *a*) Boys: 162.2 cm   Girls: 161.5 cm   *b*) Boys: 4 cm   Girls: 2.5 cm
c) 7 girls   *d*) 9 boys   *e*) 8 girls   *f*) 16 boys.

5.

	Median	Quartiles	Semi-interquartile Range
English	51	41, 59.5	9.25
Arithmetic	52.5	48, 57.5	4.75

Exercise B   Page 70

1.   2.86   2.   4   3.   5.5

Exercise C   Page 73

1. *a*) Mean = 7   ; S.D. = 2.92   *b*) Mean = 23.4; S.D. = 3.24
   c) Mean = 65   ; S.D. = 2.73   *d*) Mean = 35.5; S.D. = 2.96
   e) Mean = 29.2; S.D. = 7.60
2.   Mean = 77.7°F;   S.D. = 3.10°F   3. Mean = 15.5°F;   S.D. = 4.03°F
4.   North Scotland: mean = 19.9 days; S.D. = 1.93 days.
      East   Scotland: mean = 16.6 days; S.D. = 1.44 days.
      West Scotland: mean = 18.7 days; S.D. = 1.93 days.
5.   Mean = 163.05 cm; S.D. = 6.37 cm   6. Mean = 156.47 cm; S.D. = 5.48
8. Mean = 24.5 marks; S.D. = 8.15
9. Mean = 53.25 kg   : S.D. = 9.67 kg
11. Mean = 9.77   : S.D. = 1.71
12. Mean =   sec   ; S.D. = 4.8 sec

Exercise D   Page 78

	Mean	Standard Deviation.
1.	152.3 cm	1.82 cm
2.	53.2	1.32
3.	£83.40	£13.30
4.	24.0	9.05
5.	46.6	13.5 (First Examination)
	58.3	21.2 (Second Examination)
6.	161.8 cm	4.12 cm (Boys)
	161.6 cm	2.28 cm (Girls)
7.	50.0	14.1 (English Examination)
	52.0	7.94 (Arithmetic Examination)

Exercise E   Page 79

1. *a*) 45, 6   *b*) 30, 4   *c*) 25, 2   *d*) 20, 2   *e*) 7.5, 1   *f*) 12, 2
2. *a*) 25, 2.25   *b*) 300, 27   *c*) 90, 9   *d*) 120, 9   *e*) 205, 18   *f*) 45, 4.5
3. *a*) 115, 12.5   *b*) 24, 2.5   *c*) 20, 2.5   *d*) 11.5, 1.25   *e*) 48, 5
   f) 67, 7.5   *g*) 50, 5   *h*) 11, 1.25
4. *a*) 105, 15.6   *b*) 25, 3.9   *c*) 114, 15.6   *d*) 25, 3.9   *e*) 186, 23.4
5. *a*) 7   *b*) 2.67   *c*) 19, 5.34
6. *a*) 58.3   *b*) 5.48   *c*) 34.15, 2.74   *d*) 12.575, 1.37

Exercise F   Page 81

1.   May   2. *a*) 3rd exam   *b*) No—it was the same

3. *a*)

Pupil	Mark	Deviate from mean	Standard Score
Alan	68	68–64 = + 4	$+\frac{1}{2}$
Betty	72	72–64 = + 8	$+1$
Colin	56	56–64 = − 8	$−1$
David	60	60–64 = − 4	$−\frac{1}{2}$
Evelyn	80	80–64 = +16	$+2$
Fay	76	76–64 = +12	$+1\frac{1}{2}$

Pupil	Mark	Deviate from mean	Standard Score
Alan	58	+ 9	$+1\frac{1}{2}$
Betty	52	+ 3	$+\frac{1}{2}$
Colin	43	− 6	-1
David	49	0	0
Evelyn	64	+15	$+2\frac{1}{2}$
Fay	55	+ 6	$+1$

   c) Alan   *d*) Betty and Fay   *e*) Colin
4. Mary   5. Helen, Jean, Mary.

Exercise G   Page 81

1. *a*) 13, 2   *b*) (1) 15, 2   (2) 26, 4   (3) 28, 4   *c*) 32, 38, 44, 50, 56, 62, 68
2. *a*) 7   *b*) 7   *c*) 3.5   *d*) 12   *e*) 7   *f*) 3.14   *g*) 3.85   *h*) 19, 7.70
3. *a*) 15   *b*) 18   *c*) 11.5   *d*) 17   *e*) 4.89   *f*) 5.41   *g*) 7.5, 2.7
4. *a*) 17   *b*) 13.5   *c*) 10   *d*) 13.5   *e*) 5.59   *f*) 57.5, 27.95
5. *a*) 3 or 4   *b*) 8   *c*) 5   *d*) 5   *e*) 2.5   *f*) 6   *g*) 2.44   *h*) 2.71   *i*) 3.67, 0.90
6. *a*) 10   *b*) 65   7. *a*) 2   *b*) 4   *c*) 1.07   *d*) 7.16
8. *a*) 6   *b*) 24   *c*) 1.66   *d*) 82
9. *a*) 24.3   *b*) 6.0   *c*) 24.0
10. *a*) 23 kg   *b*) 7.5 kg   *c*) 23.4 kg   *d*) 5.73 kg
11. 6.33, 1.84, 5
12. *a*) 38.5 yr, 10.7 yr   *b*) 37 yr, 9 yr, 29.5 yr
13. 3.5, 1.40; 63.5, 1.40
14. 82.4, 12.3   15. *a*) 64   *b*) 12
16. Y, 64

Chapter 8
Exercise A   Page 87

1. Mean points (19,16), (43,26)   *a*) 29   *b*) 13   *c*) no ans.
2. Mean points $(14\frac{1}{2},15)$, $(28,28\frac{1}{2})$ *a*) 16   *b*) 25   *c*) no ans.
3. Mean points $(31\frac{1}{2},29\frac{1}{2})$, (52,39)   *a*) 32   *b*) 38   *c*) no ans.
4. Mean points (5.03,5.23), (5.7,5.53)   *a*) 5.3   *b*) 5.4
5. Mean points $(13\frac{1}{2},14\frac{1}{2})$, (29,29)   *a*) 12   *b*) 30   *c*) 18
6. Mean points (29,19), (51,34)   *a*) 16   *b*) 27   *c*) 30   *d*) no ans.
7. Mean points $(108,43\frac{1}{2})$, $(132,74\frac{1}{2})$ *a*) 50   *b*) 85   *c*) 73
8. Mean points $(46,47\frac{1}{2})$, (83,75)   *a*) 50   *b*) 65   *c*) 80
9. Mean points $(22\frac{1}{4},18\frac{3}{4})$, (55,37)   *a*) 17   *b*) 30   *c*) 39

Exercise B   Page 90

1.   *a*) $y = 2x - 6$ *b*) $y = 7 - x$

      c) $y = \frac{1}{2}x + 3\frac{1}{2}$ *d*) $y = \frac{2}{3}x + 1\frac{2}{3}$

2.   *a*) $y = \frac{3}{2}x - \frac{1}{2}$ *b*) $y = 3\frac{1}{2} - \frac{1}{4}x$

3. $y = \frac{1}{2}x + 5$;   22

4. $y = -\frac{7}{4}x + 11\frac{1}{2}$;   (1) $6\frac{1}{4}$ (2) $2\frac{3}{4}$

5. $y = \frac{1}{3}x + 8\frac{1}{3}$ 6. $y = \frac{7}{8}x + \frac{3}{8}$

7.   *a*) $y = -5x + 554$ *b*) $y = -7x + 746$
8. $y = 400 - 5t^2$ (1) 400 metres   (2) 8.9 secs   (3) 6.3 secs

9. $y = \frac{6}{x} + 4$ *a*) 4.6 pence   *b*) 1200 metres.

1. 0.94   2. 1   3. -1 4. 0.99   5. 0.89
6. -0.88 7. 0.46   8. 0.14
9. *a)* 0.33   *b)* 0.57   *c)* 0.60   *d)* 0.40

Chapter 9
Exercise A   Page **100**

1. $\frac{5}{6}$ 2. *a)* $\frac{3}{13}$ *b)* $\frac{2}{13}$ 3. *a)* $\frac{1}{6}$ *b)* $\frac{1}{3}$ *c)* 0   *d)* $\frac{1}{2}$

4. *a)* $\frac{1}{13}$ *b)* $\frac{1}{52}$ *c)* $\frac{1}{4}$ *d)* $\frac{3}{13}$ *e)* 0

5. *a)* $\frac{25}{51}$ *b)* $\frac{2}{51}$ *c)* either $\frac{4}{51}$ or $\frac{3}{51}$ depending on whether the black card removed was a king.

6. $\frac{1}{3}$ 7. $\frac{3}{4}$ 8. *a)* $\frac{1}{20}$ *b)* $\frac{85}{99}$

9. *a)* $\frac{2}{5}$ *b)* $\frac{3}{5}$ *c)* 0   *d)* 1   10. *a)* $\frac{1}{9}$ *b)* 0   *c)* $\frac{1}{12}$ *d)* 0   *e)* 1

11. $\frac{1}{4}$ 12. *a)* $\frac{1}{8}$ *b)* $\frac{1}{2}$ *c)* $\frac{1}{8}$ 13. *a)* $\frac{1}{12}$ *b)* $\frac{1}{12}$ *c)* 0

14. *a)* $\frac{1}{4}$ *b)* $\frac{1}{2}$ 15. *a)* $\frac{1}{2}$ *b)* $\frac{9}{16}$ 16. *a)* $\frac{1}{4}$ *b)* $\frac{1}{4}$

17. *a)* $\frac{1}{8}$ *b)* $\frac{1}{6}$ *c)* $\frac{1}{12}$ *d)* $\frac{1}{4}$ 18. *a)* $\frac{1}{12}$ *b)* $\frac{1}{24}$ *c)* $\frac{1}{2}$

19. $\frac{1}{16}$ 20. $\frac{1}{32}$

Exercise B   Page **102**

1. *a)* $\frac{1}{4}$ *b)* $\frac{1}{6}$ *c)* $\frac{1}{6}$

2. *a)* $\frac{1}{9}$ *b)* $\frac{4}{9}$ *c)* $\frac{1}{27}$ *d)* $\frac{8}{27}$

3. *a)* $\frac{1}{15}$ *b)* $\frac{1}{15}$ *c)* $\frac{2}{15}$ *d)* $\frac{2}{5}$

4. *a)* $\frac{5}{12}$ *b)* $\frac{3}{4}$

5. *a)* $\frac{3}{10}$ *b)* $\frac{8}{15}$ *c)* $\frac{5}{12}$ *d)* $\frac{17}{30}$

6. *a)* $\frac{1}{42}$ *b)* $\frac{9}{14}$ *c)* $\frac{9}{14}$ *d)* $\frac{5}{6}$ *e)* $\frac{11}{21}$

7. $\frac{1}{64}$

8. $\frac{1}{36}$

9. *a)* $\frac{1}{60}$ *b)* $\frac{1}{30}$ *c)* $\frac{1}{60}$ *d)* $\frac{1}{15}$ *e)* $\frac{1}{15}$

Exercise C   Page **103**
1.   50   2. *a)* 30   *b)* 60   *c)* 40
3. *a)* 20   *b)* 40   4. 80   5. 20
6. *a)* 46   *b)* 75   7. 200   8. 367
9.   80   10. 9   11. 120

Exercise D   Page 104

1.   *a)* 500   *b)* (1) $\frac{43}{500}$ (2) 0   (3) $\frac{13}{125}$

2.   *a)* 0.4   *b)* 0.36   *c)* 0.84   *d)* 36

3.   P(P) + P(Q) + P(R) cannot exceed 1.

4.   *a)* (1) $\frac{1}{9}$ (2) $\frac{8}{9}$ (3) $\frac{8}{27}$ (4) $\frac{1}{4}$

5.   *a)* (1) $\frac{343}{1000}$ (2) $\frac{657}{1000}$ *b)* 3

6.   *a)* (1) $\frac{1}{36}$ (2) $\frac{1}{18}$ (3) $\frac{1}{12}$ (4) $\frac{1}{9}$ (5) $\frac{5}{36}$
     b) (1) 3   (2) 6   (3) 8   (4) 11   (5) 14
         (6) 17   (7) 14   (8) 11   (9) 8   (10) 6   (11) 3.

Chapter 10

Exercise A   Page 110

1.   *a)* $\frac{80}{243}$ *b)* $\frac{20}{243}$ *c)* $\frac{64}{729}$ 2. $\frac{1}{81}, \frac{8}{81}, \frac{24}{81}, \frac{32}{81}, \frac{16}{81}$

3. $\frac{35}{128}$ 4.   0.6561, 0.2916, 0.0486, 0.0036, 0.0001

5. $\frac{7}{64}$ 6. $\frac{16}{81}, \frac{32}{81}, \frac{24}{81}, \frac{8}{81}, \frac{1}{81}$ 7. $\frac{459}{512}$ 8. $\frac{44}{125}$

9.   *a)* $\frac{2187}{16384}$ *b)* $\frac{2835}{16384}$ 10.   0.432   11. $\frac{53}{3125}$

12. $\frac{23}{648}$ 13. *a)* $\frac{5}{16}$ *b)* $\frac{21}{32}$ 14. *a)* $\frac{280}{2187}$ *b)* $\frac{696}{729}$

15. $\frac{7}{8}, \frac{5}{16}$ 16.   0.22, 0.40, 0.53, 0.64

17. *a)* 0.6561   *b)* 0.2916   *c)* 0.0523   *d)* 0.9999

18. $\frac{243}{3125}, \frac{810}{3125}, \frac{1080}{3125}, \frac{720}{3125}, \frac{240}{3125}, \frac{32}{3125}$ *a)* 5   *b)* 39

19. *a)* 20   *b)* 44   20. *a)* 37   *b)* 90

Chapter 11

Exercise A   Page 119

1. *a)* 1.33   *b)* -1 *c)* 0.273   *d)* -1.6
2. *a)* -0.455 *b)* 1.273   *c)* -2.455 3.   She did better in 1st exam.
4.   Jane   5.   He did equally well in all three subjects
6.   Bob   7. *a)* May   *b)* Nov.   *c)* Nov. maths   *d)* Nov. English
8. *a)* 34.2   *b)* 37.5   *c)* 25.5   *d)* 27.96   9.   3.85
10.   10   11.   21.5   12.   63.15   13. $\mu = 50; \ \sigma = 5$
14. $\mu = 65; \ \sigma = 10$

Exercise B   Page 122

1.   0.403   2.   0.492   3.   0.226   4.   0.353
5.   0.617   6.   0.953   7.   0.300   8.   0.258
9.   0.089   10.   0.441   11.   0.115   12.   0.001
13.   0.268   14.   0.081   15.   0.983   16.   0.691
17.   0.907   18.   0.633   19.   0.218   20.   0.782
21.   0.903   22.   0.189

23. *a)* 0.614 *b)* 0.242 *c)* 0.092 *d)* 0.159 *e)* 0.309
24. *a)* ±1.17 *b)* ±2.20 *c)* −0.46 *d)* 0.87 *e)* −0.24 *f)* 1.28
    g) ±2.46 *h)* ±2.75 *i)* ±2.32 or 2.33

Exercise C   Page 124

1. *a)* 0.023 *b)* 0.159 *c)* 0.081 *d)* 0.023 *e)* 0.954
2. *a)* 0.933 *b)* 0.818

Exercise D   Page 125

1.   0.004   2. 1.56   3. 0.261   4. 11.5%
5.   0.008; 0.044   6. 186.4 cm   7. 0.025   8. 40.8%
9.   77%   10. 9   11. 7 years
12. *a)* 115   *b)* 168.3 cm and 176.7 cm   13. 109.860 g and 160.445 g
14. $\mu = 22.72$; 21.04   15. 47   16. 964
17. 19.05   18. 1288   19. $\mu = 167.1$ cm $\sigma = 9.43$ cm
20. $\mu = 64.9$; $\sigma = 9.55$

Exercise E   Page 130

1. *a)* 0.029   *b)* 0.08   2. 0.112   3. 0.123
4.   0.097   5. *a)* 0.013   *b)* 0.986   6. 0.006
7. 17   8. 0.999   9. 0.002
10. *a)* 16, 3.58   *b)* 1) 0.164   2) 0.559
11. 10, 2.74, 0.100, 1 or 2
12. *a)* 2.5, 1.54 *b)* 1) 0.742   2) 0.716
13. 0.999   14. 0.031   15. 23

Chapter 12
Exercise A   Page 135

1. *a)* 0.037   *b)* 0.001   2. *a)* 0.084   *b)* 0.510
3.   0.019   4. 0.115   5. From 29.665 to 30.335
6.   £1.95   7. *a)* 96   *b)* 166   8. 240

Exercise B   Page 138

1. 30±0.96   2. 65±0.65   3. 12.6±0.74
4. 171.05±0.865 cm 5. 161.25±1.08 cm 6. 2.5±0.18

APPENDIX: TABLES

Table 1. Area under the Normal Distribution Curve

The function tabulated is

$$A(Z) = \frac{1}{\sqrt{2\pi}} \int_0^Z e^{-\frac{1}{2}t^2}\, dt$$

$A(Z)$ is the proportion of area under the normal curve between the mean ordinate at 0 and the ordinate at any value Z.

Z	A(Z)	Z	A(Z)	Z	A(Z)	Z	A(Z)
				0.60	0.226	1.05	0.353
				0.61	0.229	1.06	0.355
				0.62	0.232	1.07	0.358
				0.63	0.236	1.08	0.360
				0.64	0.239	1.09	0.362
				0.65	0.242	1.10	0.364
				0.66	0.245	1.11	0.367
				0.67	0.249	1.12	0.369
				0.68	0.252	1.13	0.371
				0.69	0.255	1.14	0.373
				0.70	0.258	1.15	0.375
				0.71	0.261	1.16	0.377
				0.72	0.264	1.17	0.379
				0.73	0.267	1.18	0.381
				0.74	0.270	1.19	0.383
0.00	0.000	0.30	0.118	0.75	0.273	1.20	0.385
0.01	0.004	0.31	0.122	0.76	0.276	1.21	0.387
0.02	0.008	0.32	0.126	0.77	0.279	1.22	0.389
0.03	0.012	0.33	0.129	0.78	0.282	1.23	0.391
0.04	0.016	0.34	0.133	0.79	0.285	1.24	0.393
0.05	0.020	0.35	0.137	0.80	0.288	1.25	0.394
0.06	0.024	0.36	0.141	0.81	0.291	1.26	0.396
0.07	0.028	0.37	0.144	0.82	0.294	1.27	0.398
0.08	0.032	0.38	0.148	0.83	0.297	1.28	0.400
0.09	0.036	0.39	0.152	0.84	0.300	1.29	0.401
0.10	0.040	0.40	0.155	0.85	0.302	1.30	0.403
0.11	0.044	0.41	0.159	0.86	0.305	1.31	0.405
0.12	0.048	0.42	0.163	0.87	0.308	1.32	0.407
0.13	0.052	0.43	0.166	0.88	0.311	1.33	0.408
0.14	0.056	0.44	0.170	0.89	0.313	1.34	0.410
0.15	0.060	0.45	0.174	0.90	0.316	1.35	0.411
0.16	0.064	0.46	0.177	0.91	0.319	1.36	0.413
0.17	0.067	0.47	0.181	0.92	0.321	1:37	0.415
0.18	0.071	0.48	0.184	0.93	0.324	1.38	0.416
0.19	0.075	0.49	0.188	0.94	0.326	1.39	0.418
0.20	0.079	0.50	0.191	0.95	0.329	1.40	0.419
0.21	0.083	0.51	0.195	0.96	0.331	1.41	0.421
0.22	0.087	0.52	0.198	0.97	0.334	1.42	0.422
0.23	0.091	0.53	0.202	0.98	0.336	1.43	0.424
0.24	0.095	0.54	0.205	0.99	0.339	1.44	0.425
0.25	0.099	0.55	0.209	1.00	0.341	1.45	0.426
0.26	0.103	0.56	0.212	1.01	0.344	1.46	0.428
0.27	0.106	0.57	0.216	1.02	0.346	1.47	0.429
0.28	0.110	0.58	0.219	1.03	0.348	1.48	0.431
0.29	0.114	0.59	0.222	1.04	0.351	1.49	0.432

Table 1 continued

Ƶ	A(Ƶ)	Ƶ	A(Ƶ)	Ƶ	A(Ƶ)	Ƶ	A(Ƶ)
1.50	0.433	1.95	0.474	2.40	0.492	2.85	0.498
1.51	0.434	1.96	0.475	2.41	0.492	2.86	0.498
1.52	0.436	1.97	0.476	2.42	0.492	2.87	0.498
1.53	0.437	1.98	0.476	2.43	0.492	2.88	0.498
1.54	0.438	1.99	0.477	2.44	0.493	2.89	0.498
1.55	0.439	2.00	0.477	2.45	0.493	2.90	0.498
1.56	0.441	2.01	0.478	2.46	0.493	2.91	0.498
1.57	0.442	2.02	0.478	2.47	0.493	2.92	0.498
1.58	0.443	2.03	0.479	2.48	0.493	2.93	0.498
1.59	0.444	2.04	0.479	2.49	0.494	2.94	0.498
1.60	0.445	2.05	0.480	2.50	0.494	2.95	0.498
1.61	0.446	2.06	0.480	2.51	0.494	2.96	0.498
1.62	0.447	2.07	0.481	2.52	0.494	2.97	0.499
1.63	0.448	2.08	0.481	2.53	0.494	2.98	0.499
1.64	0.449	2.09	0.482	2.54	0.494	2.99	0.499
1.65	0.451	2.10	0.482	2.55	0.495	3.00	0.499
1.66	0.452	2.11	0.483	2.56	0.495	3.01	0.499
1.67	0.453	2.12	0.483	2.57	0.495	3.02	0.499
1.68	0.454	2.13	0.483	2.58	0.495	3.03	0.499
1.69	0.454	2.14	0.484	2.59	0.495	3.04	0.499
1.70	0.455	2.15	0.484	2.60	0.495	3.05	0.499
1.71	0.456	2.16	0.485	2.61	0.495	3.06	0.499
1.72	0.457	2.17	0.485	2.62	0.496	3.07	0.499
1.73	0.458	2.18	0.485	2.63	0.496	3.08	0.499
1.74	0.459	2.19	0.486	2.64	0.496	3.09	0.499
1.75	0.460	2.20	0.486	2.65	0.496	3.10	0.499
1.76	0.461	2.21	0.486	2.66	0.496	3.11	0.499
1.77	0.462	2.22	0.487	2.67	0.496	3.12	0.499
1.78	0.462	2.23	0.487	2.68	0.496	3.13	0.499
1.79	0.463	2.24	0.487	2.69	0.496	3.14	0.499
1.80	0.464	2.25	0.488	2.70	0.497	3.15	0.499
1.81	0.465	2.26	0.488	2.71	0.497	3.16	0.499
1.82	0.466	2.27	0.488	2.72	0.497	3.17	0.499
1.83	0.466	2.28	0.489	2.73	0.497	3.18	0.499
1.84	0.467	2.29	0.489	2.74	0.497	3.19	0.499
1.85	0.468	2.30	0.489	2.75	0.497	3.20	0.499
1.86	0.469	2.31	0.490	2.76	0.497	3.21	0.499
1.87	0.469	2.32	0.490	2.77	0.497	3.22	0.499
1.88	0.470	2.33	0.490	2.78	0.497	3.23	0.499
1.89	0.471	2.34	0.490	2.79	0.497	3.24	0.499
1.90	0.471	2.35	0.491	2.80	0.497	3.25	0.499
1.91	0.472	2.36	0.491	2.81	0.498	3.26	0.499
1.92	0.473	2.37	0.491	2.82	0.498	3.27	0.499
1.93	0.473	2.38	0.491	2.83	0.498	3.28	0.499
1.94	0.474	2.39	0.492	2.84	0.498		

Table 2. Random numbers

```
7 1 7 1 9 8 6 3 2 6 4 5 4 5 1 6 3 0 7 6 8 4 4 0 3 0 7
  5 4 4 9 0 8 6 3 7 9 3 1 0 7 0 1 8 0 6 6 0 2 1 6 5 6
5 5 5 7 4 0 7 3 2 2 7 1 0 3 0 7 0 6 6 1 6 9 9 1 1 7 8
0 9 3 5 3 8 4 7 9 1 4 8 7 3 6 1 2 1 0 4 5 5 5 4 6 5 7
  4 2 0 2 7 2 0 1 1 4 8 1 0 4 2 6 2 1 7 2 7 2 6 8 5 9
1 9 2 7 8 9 2 6 4 8 6 0 6 3 7 3 4 6 2 6 8 4 3 6 6 0 7
  6 8 6 4 1 7 4 0 2 2 1 9 1 4 4 3 0 1 0 3 0 5 7 3 8 3 3
  4 5 7 3 7 8 4 3 5 7 3 6 4 9 5 1 1 3 8 1 3 4 0 2 9 4
6 6 6 3 5 5 0 8 8 3 6 3 4 1 0 4 8 8 6 5 3 3 2 3 9 2 1
2 7 5 3 0 0 1 6 3 9 0 0 6 4 8 0 8 6 8 3 4 7 3 7 9 0 6
  3 4 3 3 8 8 9 5 4 3 3 4 7 8 4 4 0 4 8 6 9 2 4 1 1 5
7 0 1 2 1 0 4 3 7 1 7 5 0 1 8 1 6 6 7 1 0 9 3 0 1 1 1
  6 6 6 1 6 3 8 7 0 9 2 8 3 8 7 3 9 6 1 7 9 4 7 0 1 1 7
  9 5 1 8 3 5 2 9 7 3 5 4 0 4 4 3 7 8 2 0 0 1 8 2 7 3
5 1 7 8 6 0 9 6 3 8 9 5 2 0 5 9 5 2 8 8 7 2 8 5 6 5 0
9 6 9 1 9 3 9 0 3 5 5 8 9 4 5 4 0 6 9 3 9 8 4 9 1 5 6
  2 7 6 6 2 4 4 0 4 6 3 8 3 1 0 0 7 7 0 1 2 1 4 4 0 1
1 8 4 6 7 8 0 4 3 8 9 8 3 5 1 9 7 9 9 6 9 2 6 2 0 6 1
  4 4 0 8 9 3 5 9 4 0 0 9 7 3 8 0 8 4 4 4 8 2 1 1 8 1 5
  6 7 9 1 2 6 1 5 2 7 6 8 4 7 1 8 4 3 3 3 2 3 1 2 3 0
4 2 1 5 5 4 3 7 3 7 0 7 0 5 2 7 7 1 4 0 4 4 5 0 0 2 6
0 9 9 6 4 0 7 7 3 0 8 1 9 2 6 1 5 2 5 4 3 6 6 3 9 1 8
  6 7 4 1 8 1 3 1 0 3 0 9 1 7 6 8 8 8 2 3 7 2 8 4 4 4
3 9 4 9 9 8 6 3 7 7 1 3 1 6 8 6 8 2 1 7 1 2 8 5 3 7 3
3 9 5 8 5 0 3 2 2 3 1 0 9 7 7 7 1 1 2 3 2 9 3 0 1 6 9
  2 7 6 6 2 8 6 1 0 4 3 3 7 8 1 1 4 3 2 1 0 3 9 1 3 6
1 2 2 4 9 4 9 2 3 1 2 7 5 7 9 6 3 3 6 9 5 2 1 1 3 1 2
9 3 3 7 9 0 3 9 5 3 8 7 7 8 4 1 6 5 1 3 8 5 3 2 9 0 2
  6 6 2 4 6 5 1 4 3 7 2 4 5 2 1 3 9 3 4 7 3 5 1 2 1 1
0 1 9 3 4 1 6 2 7 8 6 2 9 7 1 3 2 8 9 2 2 0 8 9 4 4 4
8 7 3 2 1 3 8 5 9 0 0 0 0 5 8 5 3 6 2 6 7 6 1 9 9 3 1
  9 9 7 2 2 4 4 0 6 1 5 8 3 9 1 5 2 9 2 9 8 4 1 4 5 9
3 2 9 4 7 2 8 1 9 2 9 9 5 1 9 6 5 7 8 5 6 4 1 1 5 6 6
  6 7 5 4 8 2 9 6 6 1 9 2 3 4 8 8 4 2 4 0 9 8 9 5 3 9 2
  8 5 9 6 4 3 5 8 5 7 6 9 8 5 8 2 1 3 2 8 4 2 2 2 2 1
3 6 2 0 9 8 0 3 1 3 5 7 8 1 2 2 2 6 4 3 5 6 6 4 8 5 9
1 5 8 0 0 8 3 2 0 4 9 1 8 8 2 0 1 9 6 5 1 3 8 8 7 2 1
  8 2 1 2 7 8 9 7 0 1 8 2 9 4 4 7 2 6 1 3 9 7 3 5 6 5
3 3 6 0 1 5 8 9 6 3 0 7 4 3 0 2 2 5 6 1 6 1 4 9 6 0 1
6 1 6 5 5 3 0 1 3 4 3 7 3 3 7 6 4 4 6 7 3 1 4 9 5 6 8
  7 4 2 5 0 6 3 8 7 2 9 6 2 8 3 0 3 9 4 2 4 5 1 7 9 9
8 3 7 5 7 0 8 4 7 1 3 9 1 1 5 3 7 5 5 1 1 8 7 5 4 7 4
7 3 0 7 2 4 8 3 9 2 2 6 5 7 8 2 3 1 0 8 7 6 3 2 0 1 5
  1 8 6 4 7 3 3 8 7 4 2 0 7 1 3 5 4 7 0 8 2 2 1 4 5 3
0 8 7 8 6 3 3 8 4 9 3 4 9 1 7 3 3 4 9 7 2 1 5 4 4 1 4
6 9 1 9 5 9 9 8 6 2 2 4 8 8 8 2 0 7 4 5 3 8 6 1 1 3 4
  8 3 6 2 2 9 8 9 1 5 0 1 3 1 7 4 3 2 7 2 3 1 7 4 3 8
1 4 8 6 4 7 1 6 4 7 5 8 7 2 1 0 7 6 1 0 3 5 5 0 3 7 1
```